Health and Safety Programs in Small and Medium-sized Businesses

This book is intended to help people newly involved in the development of workplace safety and health programs to understand principles and concepts that can help them roll out what is often referred to as a "safety program." It introduces the safety and health landscape in simple terms and provides a framework for common program start-up or program improvement action steps.

Providing small and medium-sized enterprises (SMEs) with a primer on how to think about and go about creating a safety program, this book presents a higher-level overview, framework models and suggestions, resource information, and shares concepts that can help readers not only meet regulatory requirements but also provide true value to the organization. Safety consultant Greg G. Zigulis presents readers with an accessible overview of the types of things that should be known and considered in order to create and run an effective and sustainable health and safety program.

Written for a broad professional readership, this book will be invaluable for all SME businesspersons and practitioners who need a seasoned expert to provide an understandable introduction to safety programs, whatever their situation. It is particularly relevant to start-ups and people given responsibility for workplace safety without a background in this area.

Gregory G. Zigulis has more than 40 years of safety, health, and environmental experience within different industries and is active in consulting related to program development and implementation. He is an active member of national standards development committees and advisory bodies on Occupational Health and Safety Management Systems, and is experienced in addressing occupational hazards and risks via design, redesign, and other processes.

Health and Safety Programs in Small and Medium-sized Businesses

Setting Them Up and Making Them Work Effectively

Gregory G. Zigulis

Routledge
Taylor & Francis Group
LONDON AND NEW YORK

Designed cover image: Getty Images

First published 2026
by Routledge
4 Park Square, Milton Park, Abingdon, Oxon OX14 4RN

and by Routledge
605 Third Avenue, New York, NY 10158

Routledge is an imprint of the Taylor & Francis Group, an informa business

British Library Cataloguing-in-Publication Data
A catalogue record for this book is available from the British Library

ISBN: 978-1-032-70453-1 (hbk)
ISBN: 978-1-032-70448-7 (pbk)
ISBN: 978-1-032-70458-6 (ebk)

DOI: 10.4324/9781032704586

Typeset in Bembo
by codeMantra

For my wife, Mary, who learned more about occupational safety and health than she ever signed up for. Thanks for your patience and support as I worked to give safety requirements and programs some real-world context.

Contents

1 The Value of Occupational Safety and Health Programs

The Need for and Purpose of This Book

"Safety work is today recognized as an economic necessity. It is the study of the right way to do things."[1] I've been contacted many times by people wanting to know how to create occupational safety and health (OSH) programs for their employer, or to help them know whether what they had in place was adequate. Most often, this has been tied to new plant start-ups or post-accident responses, or to intentions to obtain third-party OSH certifications.

This book is intended to help people develop or improve upon what many people refer to as a "safety program" for their organization. That might include safety-related policies, plans, sub-programs, procedures, instructions, guidance documents, forms, or other things. People use a variety of terms to describe the pieces – the documents and records that describe the organization's intended approach to managing safety. Often, the need to develop or improve upon these things is tied to audits or inspections that have identified gaps, newly identified safety issues that need correction, serious injuries or fatalities, company growth, or customer requirements.

However, unless you come from a technical safety background, or maybe even if one "does" have some background in "safety," that task might seem large. You may have questions about how to approach that; most people would. It is often unclear to those people who put new safety programs in place what occupational safety and health (OSH) regulators require. For example, you might have questions about what written programs, plans, processes, resources, and reports are required and if safety committees are mandatory. More on those topics later.

Before talking about "value," since various words are used to describe safety programs, know that the terms "safety program" and "safety plan" are commonly used interchangeably. This is in part based upon the way the terms are used within governmental regulations. I'll discuss this in more detail later. For now, know that regulators can require organizations to develop topic-specific programs or plans depending upon the type of industry and work. Organizations in turn sometimes choose to consolidate several types of these safety plans and/or programs together and refer to that compilation as their "Company Safety

DOI: 10.4324/9781032704586-1

Program," "Company Safety Manual," or something else. To avoid repeating long optional phrases, I'll use the terms "program" and "plan" interchangeably in this book, unless otherwise noted.

Beyond that, questions often arise about individual safety program pieces that, when not technically required by a safety regulator (like the Occupational Safety and Health Administration [OSHA]), might still be good to put in place to best manage safety-related requirements or expectations, and to align with other business processes.

This book will help you to work through some of those issues, and to separate what might be viewed as "hype" versus legitimate and reasonable requirements.

Frankly, there are many published sources of information "out there" about occupational safety and health, but they are often either extremely detailed and intended for full-time technical safety staff, or so extremely basic so as to tell you little of anything. The intent of this book is to share with you the kinds of things you really should know, as a starting point. I've helped set up safety programs for several types of organizations over many years now, and I've seen what works and what doesn't; what holds up and what might not be sustainable. This book shares tips regarding the "safety process" side of things and the "business" side of things.

Fundamentally, you might be wondering if your efforts will amount to anything, and if it will have been worth your time and effort to develop and roll out your program. Many of you have previously heard of the proverbial 4-inch-thick safety binder that sits on a shelf covered in dust and doing no one any good.

But that unused 4-inch-thick binder of lore does not represent what a true safety program is. A true safety program is more than a collection of documents that sit on a shelf. Where real safety programs exist, they can and do provide real value.

Some types of safety processes (parts of a safety program, explained later) can provide direct financial value. Other parts provide clear value but in ways that are difficult to measure. How much and to what extent you will obtain value for your organization will depend upon what you put in place, your company's culture, and how well your organization implements it. But the opportunity to get real safety- and non-safety-related value by implementing a safety program is truly there.

Since your organization (if it is an employer subject to OSHA rules) needs to implement certain safety programs (or plans/processes) to ensure regulatory compliance, consider that there are ways you can implement safety programs that not only help prevent injuries and illnesses, but can also help the organization in other ways. My suggestion to you is that you should implement your safety program very purposefully and carefully, knowing that there are ways you can optimize the return on your efforts. Knowing how your safety programs can provide value can help you communicate critical information to your organization and help it make the most informed decisions.

The Value in Implementing Occupational Safety and Health Programs

OSH programs implemented *the right way* provide several different types of value. In a big-picture way, OSH programs can provide risk reduction as well as other benefits. As an example, implementing an effective safety program can create environments and processes that decrease injury potential and enhance OSHA compliance, which in turn decreases operational risks. Decreasing those risks decreases the corresponding costs, which in turn means an organization can keep more of its revenue as profit.

There are several other types of value a safety program can create that people don't often have at the top of their minds, but which can be very important. These may include improving work culture, enhancing organizational reputation and increasing employee retention. It can also boost process efficiencies, expand qualifications for work, and broaden business potential.

The next several paragraphs illustrate some of these types of value in more detail.

First and foremost, there is the value of simply preventing harm, pain, and suffering to human beings that can occur as the result of employment.

Injury and Illness Prevention

The prevention and minimization of work-related injuries and illnesses is the primary goal of OSH programs. There is a significant need for this in the United States. The National Safety Council (NSC) indicates that, in 2023, there were 4,543 preventable injury-related deaths, 4.07 million "medically consulted injuries," and billions of dollars in costs.[2] Many of these injuries were incurred by workers at small and medium-sized businesses.

When it comes to "illness" prevention, I am referring to work-related illnesses. That can include things like work-related cancers, chronic irreversible diseases, heat stress, heat stroke, and hearing loss.

One would hope that injury and illness prevention is a common value held by all managers and that it should transcend other reasons for having safety programs. However, not everyone appreciates the value of defined safety programs or believes that intentional approaches to safety management help prevent injuries.

Of course, not all companies and workers have the same level of knowledge, awareness, ability to recognize hazards and degrees of risk associated with hazards, skill sets, or physical attributes. Humans have different perceptions of risk and different levels of risk that they are willing to accept. Sometimes people think that injuries will not happen to them or their employees – until they do.

Having effective safety programs in place can help address these potential causes of injury, ensure that proper processes and controls are in place, and minimize risk. OSHA believes that having injury and illness prevention programs in place helps to reduce injury rates.[3]

Compliance with Regulatory Obligations

Many types of work and processes regulated in the United States under the Department of Labor's Occupational Safety and Health Administration (federal OSHA) and state-run OSHA programs (State Plans) require written documents of one form or another (e.g., programs, plans, designations, and certifications). Some details about these written program requirements are described in Chapter 5.

There are 22 states that run their own OSHA programs covering the private sector and state and local government workers. There are an additional seven states that have plans that do not cover the private sector but do cover state and local government workers. OSHA-approved State Plans are required to be at least as effective as those of (Federal) OSHA. I sometimes use the term "OSHA" to refer to both federal OSHA and OSHA-approved State Plans; the context should make this clear.

Note that OSHA is mandated by Congress to set and enforce workplace safety standards and covers most private sector employees (meaning non-governmental employees). However, not all work organizations are regulated by OSHA. An organization must be an "employer" with one or more employees for OSHA to apply, if it is engaged in work activities that would be subject to the OSHA standards.[4] OSHA covers multiple industries, but the focus of this book is employers engaged in activities regulated under OSHA's general industry and construction regulations. Cases where other entities issue safety and health regulations include the Mine Safety and Health Administration (MSHA), which regulates mining and mineral operations of both surface and underground mining for operations under US jurisdiction. Another example is that the US Environmental Protection Agency (EPA), which has requirements that pertain to pesticides, herbicides, and other agricultural chemicals, as well as other rules for which there is some relationship and even overlap with existing OSHA regulations (such as under the Toxic Substances Control Act [TSCA] rules; more on that later).

The focus of this book is on OSHA-related OSH program requirements. We'll discuss some items of overlap more in Chapter 2 to help provide clarity of requirements.

The point being made here is that having an effective safety and health program will be of value to your organization by helping it to meet safety and health–related regulatory requirements. The failure to comply with those requirements can result in financial penalties and other negative consequences.

In addition to fundamental injury prevention and meeting regulatory obligations, however, the development and use of safety programs can provide other tangible and intangible benefits from places you might not expect.

The American Society of Safety Professionals (ASSP), a global organization dedicated to advancing the field of occupational safety and health, states that: "[The ASSP] knows from data and anecdotal information that investment in an

OSH program is a sound business strategy, for any organization regardless of size, and will lead to having a positive impact on the financial bottom line."[5]

Some examples of where value can be obtained by implementing safety programs are included below and later in this book. I'll start with some examples of value that you might not have expected.

Offering an Environment Where People Actually Like to Work

At one extreme, if you are concerned for your personal safety on the job, or have significant concerns getting hurt or killed on the job now or in the foreseeable future, will you want to go to work?

Probably not. At least, not most people. And yet, I've been asked to help organizations assess and get their level of work safety "under control" in cases where there have been several serious injuries and even deaths over the last few years. What do the employees at those locations tell their families and local communities about their employer? Do they suggest that their friends and family members should work there? Having spoken to groups of employees who worked in those environments, I can tell you that the answer is usually, "no." It can be difficult to find people who want to work for those types of organizations. (As a rhetorical question: what sort of talent can those types of companies usually attract?)

OK, so what I just described can represent an extreme case.

However, in a more typical set of circumstances, a variety of factors including the safety of the workplace and work processes will influence how people feel about going to work. It's no news or great revelation that people can love or hate their jobs. Culturally, it can be a "thing" to hate one's job and adapt in ways that are not always good; several movies have been made in that genre (for example, *Office Space*, *Clerks*, *Horrible Bosses*, and *Clockwatchers*). However, what might not be so obvious is that an organization's safety-related processes can be a part of what people enjoy! Providing a work environment where people like to work and in which they feel safe can be helpful to the business.

Increased Worker Engagement, Ability to Influence the Environment, and Support Job-related Fulfillment and Retention

It can be relatively easy to see which employers "care" about their workers based on actions and decisions they make, and which don't. Effective safety programs can provide a framework for expectations that the company has for itself and for how it will treat its workers, and what the organization commits to. Those types of organizations lay out expectations, provide resources, and help provide conditions for not only safe work, but situations where workers themselves feel safe and valued. Engaged workers, people who can show up for work and get in "the zone," can be great communicators, have great attitudes, work well with their teams, and exceed productivity expectations. They can work through situations safely and with resilience.

Safety programs that provide opportunities for meaningful participation can give workers an ability to control their work environment and allow them to provide meaningful suggestions and other contributions. Some safety program elements can be made fun and help create camaraderie. It's well known that those types of circumstances can contribute to feelings of well-being and reduce work-related stress.

Putting safety-related processes in place that provide those types of opportunities can help both workers and the organization. What could that be worth to *your* organization?

Examples of safety program elements that can create engagement, hazard control, and satisfaction are described in later chapters.

Now we'll shift to the discussion of other types of value.

Direct Cost Avoidance: Reduced Program Costs

Work-related injuries and illnesses obviously hurt our workers and harm their families. They affect workers financially because even with workers' compensation programs they are never really made whole. They also affect the organization financially – both directly and indirectly. Avoiding injuries and illnesses will result in avoiding those harmful issues.

Some people still believe that injuries and illnesses are just a cost of doing business – that it is why the organization carries insurance. However, that is really a backward and passive way of thinking about worker safety; it misses the potential opportunities in having safety programs in place. People who say that tend to not understand the full range of potential costs – or the opportunities that can be created by starting programs to minimize them. Relying on insurance coverage is essentially waiting for bad things to happen, which could have been avoided in the first place. As the saying goes, "Why would you do that?" Employers don't get workers' compensation and other types of insurance policies for free.

Examples of direct costs that can be avoided or minimized through injury prevention programs (safety programs) include the following:

- Medical treatment costs, short- and long-term, including medical professionals, rooms, tests, and prescriptions
- Workers' compensation case administration fees
- Payment to workers for time off work
- Civil liability costs
- Lawsuit-related costs (attorneys' fees, experts' fees, court costs, etc.)
- Property losses

Remember, when all is said and done, *employers are paying these costs.*

Indirect Cost Avoidance

Effective safety management can minimize the indirect costs associated with processes impacted by worker injuries. Otherwise, when a worker is injured, the company may incur costs because of the need to:

- Reschedule production or maintenance tasks, which can in turn decrease productivity
- Retrain the injured person and possibly other staff (if appropriate and inadequate training was a cause of the incident)
- Pay overtime to those who must take over the duties of the injured worker
- Hire additional workers
- Train persons who need to take over for the injured worker
- Delay projects that could otherwise be making money
- Investigate the incident and take corrective action

The National Safety Council has estimated that the indirect costs of injuries can be up to ten times higher than direct costs.[6]

Other indirect costs and risks can relate to the facts that:

- Injuries can reduce the output of whatever your organization's product is. This in turn can reduce revenue.
- For self-insured companies, higher rates of injuries can result in the need to increase the cash reserves available to pay for claims. This can amount to the loss of access to hundreds of thousands or millions of dollars that could have otherwise been put to use for other business processes.
- Serious worker injuries and fatalities can result in negative publicity. That can in turn result in a loss of business and customer confidence, depending on the size and type of business.
- Higher rates of injuries and related costs can result in higher discretionary and administrative fees from your insurance company.
- Some injuries could result in the unexpected need to spend money to fix violations, make improvements and pay penalties outside of planned budget cycles. If that happens, there is an impact on the organization's current assets and cash flow, besides the fact that the need to manage in crisis mode is less than fun.

Injuries are unfortunate for the worker and are almost always preventable. OSHA offers a couple of tools to help demonstrate how injuries and illnesses can impact your company, as part of OSHA's Safety Pays program. The tools are located at https://www.osha.gov/safetypays/. The tools utilize average claim costs provided by the National Council on Compensation Insurance (NCCI) for cases that had involved lost time workers' compensation claims.

Increases in Efficiency in Other Areas of the Business

Organizations that understand "safety" know there is a relationship between safety-related processes and other business processes. As the old saying goes, "everything is related to everything else." It is possible to both create circumstances and processes needed for safe work, and to also obtain improvements to profit centers of the business.

For example, there are cases where the evaluation of line-side ergonomic risk factors identified hazards that were fixed at low cost, that not only reduced injury potential but also reduced manufacturing cycle times. The result was that the sites reduced likely future injury costs while also increasing rates of production, for a minimal investment.

Another example is that of machines cutting metal parts within an enclosure, which generated chemical mist. Originally, workers were exposed to the mist after opening the enclosure's doors to remove and inspect the parts that were being machined. Additional exhaust ventilation was added to evacuate the chamber better before employees needed to open the doors. After the change, the mist levels were found to be lower than allowable exposure limits. Workers felt better about being in the area. The workers were able to remove, inspect, and set up parts more quickly to start the next cycle. This resulted in air-related compliance, made the workers more comfortable in their work environment, and increased productivity.

Prerequisites to Winning Work and Increased Business Opportunities

There are many cases where an organization is required by its customers to have safety and health programs in place. Customers have varying expectations to include some that can exceed those of OSH regulators. I've been contacted by many organizations over time that had expanded in size or scope of their offerings and found themselves facing a higher level of customer expectations.

This is especially the case in the construction industry. In some cases, if your firm is experiencing injuries, your company's work may be shut down until you fix your problems. In construction, no work can mean no pay.

Many construction project owners require their construction managers and lower tier contractors to be "pre-vetted" and found to have certain safety programs in place, and to have performed (at least injury-wise) adequately over time. In these cases, even smaller contractors can be expected to have programs in place. If what the company has does not meet vetting expectations, the company may not be eligible to bid for work. This pre-vetting or pre-qualification process exists within many organizations' purchasing and engineering groups.

If your customers require your company to have safety programs in place to win or retain work, there is obvious value in your company being able to play in the game. Beyond that, however, having a solid safety program in place can get you much more than just an entry slip.

Financial Savings from OSH-related Improvement Projects

Your organization may have used tools and methodologies intended to improve productivity, performance, and efficiency. For example, it may have applied Lean Production, 5S (or 6S), Six Sigma, Gemba walks, or other techniques. These can be great tools, but organizations do not always use them to evaluate and improve safety-related processes.

What you should know is that some organizations have structured their safety programs to require the evaluation of safety-related processes when applying improvement tools like these. Doing so can and has identified improvement opportunities that result in significant cost savings or avoidance.

Requiring the use of these tools results in critical thinking that can identify change opportunities that will enable positive cost savings and/or avoidance. I am aware of many different types of improvements that have been created. Some of these opportunities required minimal effort and nominal expense. Many identified changes were of a procedural nature and did not require "investment" so much as changes in approach or methods. Some projects did require capital investment and were funded after meeting organizational investment criteria (e.g., positive net present value [NPV], acceptable anticipated returns on investment [ROI], and acceptable payback periods [PBB]). After implementation and validation and over time, these projects were shown to provide major value.

At this point, I've described many types of value that can be obtained by implementing effective safety programs and given some examples. Since organizations regulated under OSHA (or OSHA-approved State Plans) must by law have at least certain safety programs in place, why not design those safety programs to your advantage? You can and should. We discuss this in more detail in later chapters.

Let's move now to a discussion of where requirements for an organization to fulfill occupational safety and health requirements like safety programs or plans can come from, and look at some additional context.

Notes

1 This quote from Robert W. Campbell, first president of the National Safety Council, is an excerpt from the Proceedings of the National Safety Council Third Annual Safety Congress, October 13, 1914. It was taken from the National Safety Council's website, nsc.org. The point of including this quote is to illustrate that a relationship between "safety" and "economics" has long been understood. Concepts about "safety" and what it takes to "be safe" or "provide a safe workplace" have changed over time; some of these factors are discussed in later chapters.

2 Data for 2023 were extracted from NSC Injury Facts, injuryfacts.nsc.org/work/work-overview/work-safety-introduction/, on February 26, 2025. The costs of injuries from 2023 were not yet available at the time this information was obtained from NSC, but had amounted to $167 billion dollars in 2022 (as retrieved July 4, 2024).

3 U.S. Department of Labor, Occupational Safety and Health Administration. (2012). Injury and Illness Prevention Programs White Paper. https://www.osha.gov/sites/default/files/OSHAwhite-paper-january2012sm.pdf.

4 There are some exclusions. For more detailed information on this precise point, see the Department of Labor's Employment Law Guide at https://webapps.dol.gov/elaws/elg/osha.htm. This provides details on OSHA's applicability and inapplicability.

5 ASSP White Paper on the Return on Investment for Safety, Health and Environmental (OSH) Management Programs, ASSP, Approved by Approved by the Council on Practices and Standards and the ASSP Board of Directors, June 8, 2002, Reaffirmed/Reviewed June 2008, June 2010, June 2017, and June 2019.

6 NSC. (2013). Journey to safety excellence: The business case for investment in safety – A guide for executives. https://www.nsc.org/getmedia/d81515ce-57ba-4347–821e-4af731076260/journey-to-safety-excellence-safety-business-case-executives.pdf?srsltid=AfmBOorjy-0uNh--GS9sDaKiRyxRPNVHbNTUT1My7q8q7E3k_nGqtlxR.

2 Sources of Requirements

Why Do Organizations Need Occupational Safety and Health Programs?

Is it simply because of OSHA?

Many people who are newer to the occupational safety and health "arena" think that the need to have occupational safety and health (OSH) programs is mostly or entirely driven by Federal Occupational Safety and Health Administration (OSHA) and/or State Plan programs. With OSHA's press releases, information in the media about OSHA investigations of fatalities and serious injuries, and frequent references in other contexts, it could be easy to think so. OSHA requirements have played a central role in how organizations in the United States have approached safety and health since OSHA began operations in 1971.

However, OSHA's (and State Plans') requirements to have written and other programs are not the only sources of requirements for safety plans. There are many other drivers for written programs and for organizations to find "some way" to define how they want safety to be approached. When you develop your safety program, you should make yourself aware of when OSHA requires written programs or plans, but also that OSHA is not the only source of safety-related requirements.

Some organizations' leaders have an inherent desire to look after their organization and to keep people "safe." Many of those types of leaders simply know and believe that this is the right thing to do. Others may have come to the point that they feel that way because they've known people who were killed on the job or seriously and permanently injured. That kind of experience often leaves a mental scar; persons in that position seem to take safety very personally and take action to make sure that similar things will not happen again.

Yet others interpret levels of risk and necessary levels of involvement in another way. They may be "concerned" for safety but their frame of reference may be more strongly aligned with understanding linkages between safety and other business outcomes. Managers with this mental mindset sometimes simply want to fulfill their obligations and to be responsive to things that drive or influence business success. Recall from Chapter 1 our discussion of the value having

DOI: 10.4324/9781032704586-2

safety programs can bring to an organization, beyond helping to achieve a state of regulatory compliance.

There are other drivers that motivate organizations to have general or specific types of safety programs. These can be contracts, company qualification and vetting, general business management, broader risk management to include for third-party liability, and other things. When programs are developed, there are factors to consider in determining what should be included and what should be written versus what can or perhaps should be unwritten. As you probably know or suspect, there are also many factors that influence a program's effectiveness.

Let's discuss first the regulatory framework for why organizations may need some semblance of occupational safety and health (OSH) programs or safety plans.

Regulatory Drivers

Where exactly do the regulatory requirements for employers to create safety and health programs come from?

While OSHA is mandated by Congress to set and enforce workplace safety standards and covers most private sector (non-governmental) employees, most but not all private sector work organizations are regulated by OSHA. For example, the Mine Safety and Health Administration (MSHA) regulates mining and mineral operations of both surface and underground mining, except for certain non-mining operations and facilities on mine sites. (For a fuller description, see the agreement between MSHA and OSHA at https://www.osha.gov/laws-regs/mou/1979-03-29.[1])

Also, federal OSHA regulations do not apply to state and local government workers. (There are, however, state programs that cover the public sector.)

Note that various federal, state, county, and municipal governments may set standards of one form or another that could create worker safety and health implications. For example, the US Environmental Protection Agency (EPA) sets some health and safety–related controls and exposure limits under the Federal Insecticide, Fungicide, and Rodenticide Act (FIFRA) under its Worker Protection Standards and the Toxic Substances Control Act (TSCA).[2] FIFRA sets requirements under its Workplace Chemical Protection Program (e.g., pesticide applicators' licenses and training requirements). EPA sets some exposure limits under TSCA. Some EPA regulations can overlap with existing OSHA standards, such as for EPA's Risk Management Program (RMP) requirements pertaining to chemical accident prevention.

For some additional information, even though it is old, you may find the Centers for Disease Control (CDC)/National Institute for Occupational Safety and Health (NIOSH) publication *Making Sense of Regulations*[3] informative.

The reader should know that where other regulatory bodies may create requirements, those other requirements may or may not be able to be blended (from a program development standpoint) on a case-by-case basis. Sometimes the requirements of non-OSHA regulators and permits can be blended or combined with OSHA-driven requirements, but not always.

OSHA Regulations and Standards

When the Occupational Safety and Health Act of 1970 was passed (29 U.S.C. §651 et seq. [1970]), it created the Occupational Safety and Health Administration, also referred to as OSHA. The Act can be found on the osha.gov website, along with OSHA regulations, topical and sector information, information about Federal and State programs, OSHA publications, news, and other resources.

We probably all know that the purpose of OSHA was and is "to ensure that employees work in a safe and healthful environment by setting and enforcing standards, and by providing training, outreach, education and assistance."[4]

The Act gave OSHA only a limited period of time to promulgate its first set of standards. OSHA was allowed to, and did, rely upon selected consensus and proprietary standards as they existed at that time, along with some federal laws already in effect, to develop that first set. (This is different from how OSHA was required to promulgate standards after that – a much lengthier and more involved process.)

OSHA issues different types of rules. OSHA "standards" like the hazard communication standard at 29 CFR 1910.1200 or the fall protection standard at 29 CFR 1926 Subpart M are about specific things that employers must do in order to keep workplaces safe. OSHA defines standards as pertaining to "conditions, or the adoption or use of one or more practices, means operations, or processes, reasonably necessary or appropriate to provide safe or healthful employment and places of employment" (29 CFR 1910.2).

However, OSHA also issues rules of an administrative nature that employers must do (as applicable), such as maintain injury and illness records. All of these requirements are published in the Code of Federal Regulations (CFRs).

The terms "regulation" and "standard" are often used interchangeably. OSHA itself says that a standard is a regulatory requirement that is used to determine whether employers are in compliance with the OSH Act laws.

Clear as mud? When work is performed that is covered by the OSHA standards published in the CFR, OSHA's requirements are enforceable. Employers also must comply with the requirements of the General Duty Clause in the OSH Act (explained in the next section).

As to when work is "covered" by an OSHA standard, the OSHA standards often have a clear "scope" or "scope and applicability" section. If not, a little more reading may be necessary to determine the scope. Just read carefully. Sometimes it can be easy to think that a standard pertains to you that really doesn't, based on a particular scope.

Some OSHA regulations and standards have evolved over time, while others have remained largely the same since their inception.

It would not be feasible to provide a state-by-state summary in this book, since regulations change and state requirements can vary. This chapter is intended to be about the "why" for OSH programs, as opposed to the "what." We'll talk about more of the "what" later. If you would like to take a deeper dive into OSHA safety concepts, consider attending an OSHA 30-Hour Outreach Training[5] ("OSHA 30") covering either the general industry standards or the construction industry standards, the OSHA 510 course (Occupational Safety and Health Standards for the Construction Industry), or the OSHA 511 (Occupational Safety and Health Standards for General Industry). Beyond that, courses can be found that review specific standards or types of hazards more deeply through Outreach Training or other providers (see Chapters 4, 9 and 12).

An OSHA Approved State Plan (State Plan) can have requirements that exceed what federal OSHA requires, unless a given state has its own law that prohibits its Plan from enforcing regulations that are more stringent than those set by federal OSHA. This is currently the case in Kentucky. Therefore, when developing a written program, one needs to know which regulations (federal or state) apply. If your company works in a multistate area, it might have some locations governed by the federal regulations, and others by State Plans.

We will still discuss some standard OSHA-driven written program requirements in Chapter 5.

Vertical and Horizontal Standards

Wait, what? There's "that?" Yes. But think about it from an OSHA perspective. If you were writing regulations, there might be some things that apply across the board, and yet there might be some things you'd like to regulate that only apply to certain types of work right?

So, yes, the OSHA standards we most often think of can be considered either "vertical" standards or "horizontal" standards. These are defined in 29 CFR 1910.5(c). You can look this up on the OSHA website by doing the following:

- Click on "Standards."
- Click on "Laws & Regulations."
- Scroll down to the header of "Standards" (again).
- Click on "General Industry."
- Click on "1910.5 – Applicability of standards."
- Scroll down to section 1910.5(c).

Horizontal standards are standards that apply across multiple industries. An example from federal OSHA regulations is the Hazard Communication standard, located at 29 CFR 1910.1200. Vertical standards are standards that are focused on a particular industry (such as the construction industry) or operations, processes, equipment, or other particular things. Examples of vertical standards (beyond the construction standards just mentioned) include those for scaffolding and fall protection. (Different industries and types of work have different requirements.) By the way, vertical standards take precedence over horizontal standards, and OSHA has rules about interrelationships.

The General Duty Clause

Now, as a separate concept, there can be hazards in a workplace that no specific OSHA horizontal or vertical standard addresses, right? Even so, employers have obligations for safety in the workplace. This is described under what is sometimes called the General Duty Clause of the Occupational Safety and Health Act.

Duties of the employer and employee under the OSH Act are, from Section 5 of the Act:

> (a)
> Each employer –
> (1)
> shall furnish to each of his employees employment and a place of employment which are free from recognized hazards that are causing or are likely to cause death or serious physical harm to his employees;
> (2)
> shall comply with occupational safety and health standards promulgated under this Act.
> (b)
> Each employee shall comply with occupational safety and health standards and all rules, regulations, and orders issued pursuant to this Act which are applicable to his own actions and conduct.

OSHA can cite employers for safety hazards under that General Duty Clause in some cases. In order to do so, there must be (or have been) a serious recognized hazard (likely to cause death or serious physical harm), for which an employer failed to take action to prevent or abate. Also, the hazard had to have been "recognized," and a feasible and useful method had to have been available to the employer to be able to abate the hazard. To be "recognized" does not necessarily mean that the employer itself was aware of a hazard in a particular circumstance. For example, recognition could also stem from "common sense recognition" or recognition from within the industry in which an organization works.

There are several ways that organizations can potentially learn of hazards that it can control. This can include through industry and association publications, other company facilities with similar processes, safety-related consensus standards, through experiencing injuries of a similar nature previously, and other ways. One of those other ways is to conduct inspections and by conducting what are often called hazard analyses (or job hazard analyses) and other types of hazard and risk assessment processes. This is discussed in Chapters 4 and 5.

Note that noncompliance with requirements of OSHA's regulations can result in citations for violations of its regulatory requirements, along with penalties of various classifications and dollar values. When employers walk down that path too many times with significant violations, they may get labeled as severe violators, and at some point possibly be subject to negotiated settlement agreements. Those can mandate certain costly and disruptive activities or approaches. Agreements could pertain to an individual location or be corporate-wide (enterprise settlement agreement), where certain conditions are met. It makes sense to take action to stay off OSHA's severe violators list.

Consensus Standards Are Different than Regulatory Standards

One of the things that can be confusing to people first taking on safety responsibilities is how *consensus* standards relate to *regulatory* standards. This can be in part because uninformed individuals and articles in the press or sales literature sometimes imply that consensus standards always represent legal, mandatory compliance documents. It can get confusing, but – no, they are not always interchangeable.

OSHA standards are OSHA standards: legal requirements if you are working in a covered industry and performing certain types of work or have certain types of processes or use certain types and volumes of hazardous materials or do other things that are included in the scope and applicability of coverage.

Consensus standards on the other hand are developed via other organizations or standards-setting bodies. This is done through the participation of various experts and types of organizations. The development processes follow a particular protocol to achieve substantial agreement on the content of the standard. They can be developed for a variety of topics that could include guidelines for management, product specifications, approaches to machine safety, engineering, and others.

They can only "sometimes" carry the weight of the law; that is, when the consensus standard is specifically built into (referenced in) the OSHA standard. This is referred to as "incorporation by reference" (IBR). Note that when standards are incorporated, the addition is limited to the section(s) and publication date of the consensus standard. Therefore, just because a consensus standard is updated, it does not mean that the updates will necessarily be incorporated into the OSHA standard. The IBR list may be found at 29 CFR 1910.6.

So, if consensus standards do not always contain legal requirements, what is their value? Isn't it just a lot of extra work to look at them?

Not necessarily. OSHA standards exist for many subjects, but not for all the hazards that are addressed in consensus standards. OSHA standards can seem to vary in content from seeming to not have much detail to having a lot of detail that could also be confusing. They are not necessarily written in a way intended to give a reader context on an issue or provide background or details underlying the specific requirements.

Additionally, OSHA can be slow to update its standards. The rule-making process can take years to work through and be filled with lawsuits and politics.

On the other hand, consensus standards can be updated more frequently and be kept more current with technological developments and advancements in approaches. They can provide helpful background that might never be included in a regulation due to length or other issues associated with the rule-making process.

Consensus standards can be very helpful references. They are compiled and have been approved by professionals who work on committees with the desire to share helpful information. Sometimes that might be most helpful to, say, a manufacturer. However, many times they are prepared with safety systems and the worker in mind, sharing helpful and sometimes state-of-the-art information and best practices. Many can be useful to people developing a safety program. When you are developing programs or even project plans, it can make sense to scan consensus standards for ideas.

Consensus standards can also be used for organizational awareness and decision-making processes. For example, I've had circumstances where I've felt that a particular approach to safety management made sense, but the approach was not clearly mandated by OSHA and met with some management resistance. I've used non-IBR consensus standards in several cases to provide a frame of reference, provide some perspective of options, and demonstrate that the approach was consistent with a relevant standard that was recommended by industry experts.

Conversely and to stay balanced, some consensus standards can be written with viewpoints or approaches that would not be feasible for certain types of organizations, or contain significant information that has no relevance to a particular type or size of organization.

Readers should, in my opinion, view consensus standards as containing information that has sometimes but not always been incorporated into law and can potentially be useful. They can be obtained and reviewed for purposes of standardization and ideas for improved practices, technological advances, and other helpful information.

Finding the (often old) version of standards that are currently IBR is not always easy and can take some research. They may be found at some locations for free. Sometimes consensus standards may be found online in read-only mode for free, such as through the National Fire Protection Association (NFPA). And yet others are available for purchase or through online subscription services.

Many people are not familiar with the potential use and range of available consensus standards. As of the writing of this book, the American National Standards Institute (ANSI) website indicated that there are 14,070 American National Standards. A few examples that you might want to be aware of include the following (all are ANSI/ American Society of Safety Professionals [ASSP] standards or series of standards):

- A10: Construction and Demolition Operations series of standards
  - This includes a standard on *heat stress management* in construction and demolition operations

- A1264.1: Safety Requirements for Workplace Walking/Working Surfaces and Their Access
- A1264.2: Provision of Slip Resistance on Walking/Working Surfaces
- Z117.1: Safety Requirements for Entering Confined Spaces
- Z10: Occupational Health and Safety Management Systems
- Z15: Safe Practices for Motor Vehicle Operations
- Z16: Safety and Health Metrics and Performance Measures
- Z244: The Control of Hazardous Energy – Lockout, Tagout, and Alternative Methods
- Z359: Fall Protection Code series, including Safety Requirements for Self-Retracting Devices
- Z490: Criteria for Acceptable Practices in Safety, Health and Environmental Training
- Z590.3: Prevention through Design – Guidelines for Addressing Occupational Hazards and Risks in Design and Redesign Processes
- Z310: Risk Management and Assessment (under development at the time of writing this book)

Other important consensus standards include (but are not limited to):

- ANSI B11 series: Related to the safety of machinery
  - This includes the B11.0 – Safety of Machinery, B11.19 – Performance Requirements for Risk Reduction Measures (Safeguarding) and others
- ISO 31000: Risk Management – Guidelines
- ISO 45001: Occupational Health and Safety Management Systems
- NFPA 70E: Standard for Electrical Safety in the Workplace
- NFPA 30: Flammable and Combustible Liquids Code

The standards mentioned above are a few examples out of many that exist. For that matter, from the list above, some are part of a series of standards. In those cases, there can be many subtopics and underlying standards or guidance documents that you may not guess exist unless you investigate the series. For example, there are currently 27 core standards in the ANSI B11 series, and numerous related technical reports.

Examples of a few other guidance documents and technical reports include:

- ASSP GM-Z10.100–2024: Guidance and Implementation Manual for ANSI/ASSP Z10.0–2019 Occupational Health and Safety Management Systems
- ASSP GM-Z10.101–2019: Guidance Manual: Keeping Your People Safe in Smaller Organizations
- ASSP TR-31010-2020 Risk Management – Techniques for Safety Practitioners

Not all consensus standards apply to all organization types. Since there are so many standards that exist, you may want to conduct a search for those that are or could be applicable to your organization.

So – What Is Meant by the Term "Safety Program," Anyway?

Company Perspective

As mentioned earlier, there are many ways that people can describe an organization's approach to safety. Organizations can choose what they want to call their written documents. There is no single right way, except that sometimes the names can make it easier for a regulator or others to understand what the organization is trying to do.

Some might choose to call their approach a program, a plan, or something else (OSHA uses both terms in its regulations). Organizations usually have procedures as well to describe stepwise actions required to accomplish different types of work; these supplement plans and programs. Plans, programs, and procedures put together in a synchronized and interrelated way might be called a safety management system (depending on whether they are intended to be linked). So, what title or name is given can depend upon the industry, a parent organization, or what an organization thinks their regulators want to hear.

An organization's macro approach to safety is often called a safety program or plan and may be contained in a "safety manual." From an organization's standpoint, when people talk about a safety program or plan, they are often talking about a set of documents that together describe the organization's overall approach to safety.

Organizations will often prepare that set of written documents in a tiered format and include several "kinds" of documents within it. The exact structure often depends on the industry, company, and type of organization. For example, one approach may work best for manufacturing, another for field services, and yet another for construction. The organization's approach may be to include a set of interrelated documents that help it to meet regulatory requirements but also describe how it wants safety functions to be performed. For example, they may include an overarching policy, but also critical procedures ("standard operating procedures" or "safe work procedures"), work instructions, and forms. They may also describe things that are not directly required by OSHA but that the organization feels are important enough to call out. That can include, for example, goal-setting processes and internal/external auditing requirements (beyond inspections required by OSHA). Sometimes safety committees are required by regulation, but not always. Either way, if there is a safety committee or team, an organization may have a "safety charter" that defines its purpose, membership, and how it is intended to run.

Depending on the size and complexity of an organization, company programs will sometimes describe linkages between related functions. For example, training programs may be linked if they are not administered by "Safety." Another example is that a safety program may describe interrelationships or links to processes intended to prevent workplace violence, or those addressing environmental management. Still others might describe how safety processes support other multifunctional goals or processes such as for operational excellence or lean Six Sigma–based improvement processes.

OSHA Perspective

However, when OSHA requires something written, it sometimes refers to *those* written "somethings" as plans, or as programs. It sometimes also requires written procedures, as something to supplement or support the written programs or plans.

So, in OSHA-speak, OSHA can require an employer to develop a very specific program to meet a very particular standard. For example, if an organization is subject to the Hazard Communication Standard, then the OSHA standard requires it to develop a written hazard communication *program*. For those requiring the use of respirators, a written respiratory protection *program* (with limited exception). However, the bloodborne pathogens standard at 29 CFR 1910.1030 (for those subject to it) requires the development of a written exposure control *plan* (although also a training program). Other regulations describe requirements for emergency action *plans* and chemical hygiene *plans*. Possibly creating extra confusion is the fact that for some topics federal OSHA requires programs but does not explicitly require that the program be written (it depends).

Sometimes OSHA requires both written programs/plans and procedures. Procedures are usually step-by-step process descriptions that describe steps or process flow to know "how" the organization expects people to get things done.

From a practical standpoint, organizations need to determine how they want to take OSHA's requirements for written and unwritten programs and fit them into whatever macro-framework (my term) the organization chooses to use. The organization needs to ensure that the macro-program includes the programs, plans, and procedures required by OSHA, and that they are written to include all the topics and elements that OSHA requires.

Other Drivers and Influences (Going Beyond Pure OSHA!)

So far, we've reviewed information about perhaps the largest driver of safety programs, the OSHA regulations. We've also discussed how regulatory standards relate to consensus standards, and the role and importance of consensus standards. And we surveyed some often-used terminology associated with organizational OSH programs.

However, there are many other potential sources of requirements or reasons for organizations to develop or have OSH programs. They are not always immediately obvious and yet are important to consider. They could influence how you think about developing your program.

Figure 2.1 Examples of Primary Drivers for Occupational Safety and Health Programs

For the sake of discussion and not anything official or stemming from OSHA, I've categorized other sources of requirements into "primary drivers" and "other important influences." The classification is mostly relative and depends on industry.

Let's review them; please see Figure 2.1 and the text that follows.

Other Primary Drivers

(Other [non-OSHA]) Federal and State Government

Other federal agencies can have rules that impact how your organization handles its safety obligations. This includes the Department of Transportation (DOT) and the Environmental Protection Agency (EPA).

The EPA has many types of requirements that impact safety, and requirements for approaches to safety that can interact with the balance of your OSHA-driven programs. This is especially the case regarding the Toxic Substances and Control Act (TSCA) rules, which set Existing Chemical Exposure Limits (ECELs) (in contrast to OSHA's Permissible Exposure Limits [PELs]) and Workplace Chemical Protection Programs (WCPPs).

There are sometimes requirements for plans contained in environmental-related regulations that can overlap with other OSHA-driven requirements.

Note also that, regarding environmental regulations, depending upon where you operate and the type of environmental matter in question, you may be subject to regulations from and enforcement by federal programs, state programs, or both. This is in addition to possible local requirements or agreements.

You should make sure you know about regulations of other agencies or entities that may or may not align with OSHA's requirements.

Also, some states have safety-related requirements that are promulgated through state codes that are not part of a state OSHA program, e.g., fire codes with requirements about fire extinguishers, evacuation plans, and emergency exits.

Local Government

Some local government entities can have extra requirements for training, attendance at safety meetings, enhanced scaffolding or fall protection requirements, special inspections, and other things that go beyond federal OSHA or State Plan requirements. Some can have special licensing requirements depending upon the type of work or equipment used. That can include, for example, local licensures for crane operators, and permitting for scaffolding. Some municipalities require, on a mandatory basis, the completion of an OSHA 10-hour or OSHA 30-hour class to be able to work on projects that are of a particular type, depending on job role.

Workers' Compensation

Most employers are required to maintain workers' compensation insurance, although there are some exceptions.

Some workers' compensation laws require employers to have a safety plan or program. Some require that employers seeking approval to self-insure have a written program.

Also, some workers' compensation laws require that organizations have safety committees under certain conditions.

Additionally, workers' compensation laws require that employers report injuries under varying criteria.

The bottom line is that it makes sense to be aware of what the workers' compensation–driven requirements are for the locations at which your organization operates.

Other Liabilities

There are laws that organizations can be subject to, apart from OSHA's requirements and the requirements of other agencies. Examples include obligations relative to third parties, wrongful death, and the claims by an employee of a subcontractor against a general contractor or prime contractor. Even though your organization may have workers' compensation coverage and with limits on compensation to injured or ill employees, if your organization is sued in some other context, it could lose enough money to impact its continued viability. It can make sense for organizations to consider these exposures and ensure that controls or mitigations are in place.

Financial/Insurance Requirements

Insurance companies may require that a company have certain safety programs in place to issue or renew certain types of coverage.

Loss control/risk management services from brokerages and insurance companies will also visit companies and operating locations to assess risk and make recommendations for safety-related improvements. I've seen that most consistently with property coverage. Even "property" coverage has human safety aspects; a fire cannot only ruin facilities and processes and disrupt business but also injure and kill people. Sometimes other loss control services might be included as part of your insurance package. For example, a brokerage may be able to supply services to assist with ergonomics, industrial hygiene, electrical safety, or other needs.

Organizations receive reports from these visits and are asked to respond. Reports can include both regulatory and best practice suggestions. The bottom line is that risk drives cost.

Owner/Parent/Corporate Requirements

I'm aware of several cases where a corporation's board members learned of fatalities in an organization or other adverse trends and required it to implement improvement programs.

More commonly, however, organizations of a particular size will have OSH requirements that it will require of ("flow to") new locations, branches, subsidiaries, etc. What flows down may describe overall requirements for a business unit or location, but can sometimes be very detailed.

Often when a company grows through acquisition, before it acquires a new business, it will conduct a due diligence review of OSH and environmental programs. Improvements may be needed as a condition for purchase, or liabilities otherwise addressed in negotiation. Sometimes a newly acquired location will be asked to improve or revise their safety programs in order to meet the parent organization's standards, within a given period of time. (I'm also aware of, unfortunately, circumstances where the acquiring organization had programs and expectations that were not as rigorous as the company it purchased!)

Unfortunately, sometimes it is not until an organization is up for sale that it improves its OSH and environmental profile to make it more attractive or to meet conditions for sale. That represents a lot of lost opportunities over time!

Employees/Workers/Volunteers

Of course, most workers care about their personal safety on the job – especially now. Workers usually don't expect to be in one job forever, and worker loyalty seems lower. The median tenure for workers was recently reported to be at 3.5 years for private sector employees and 6.2 years for public sector employees.[6] Along with lower job tenure, the mix of generations in the workplace of course keeps changing. While each person is unique, workers place greater value on the meaningfulness of their work and seem to have a higher willingness to "work elsewhere" if expectations don't align.

It does not take long for workers to draw conclusions about whether an organization is "good" about supporting worker safety, and when not so good. When workers see that an organization cares about safety through not just words but also actions, there are positive benefits. These can include the development of a positive safety climate and longer-term culture. That can both keep workers safe and contribute to the belief that the company is good to work for. It can also reduce organizational risk and achieve other operational benefits.

On the other hand, we probably all know that where there are perceptions or where there is the reality that a company does not care about worker safety or put in much effort toward it, problems often follow. For some people, that can be demotivating. For others, it can send a signal that the "end justifies the means" and result in higher levels of noncompliance with safety rules and of risk-taking behavior.

Business Vetting/Qualifications

Businesses sometimes need to meet certain safety criteria in order to be able to qualify to be a supplier to another organization or qualify to work as a contractor to another organization.

In some cases, an organization may be able to simply describe their safety program and performance to a potential customer initially or periodically, in order to be considered eligible for selection or re-selection as a vendor.

In other cases, however, organizations must be vetted by third parties to be deemed eligible to even bid for work. That vetting process is big business. Many operate in that space, including Avetta, ISNetworld (ISN), Highwire, and others. Contractors must submit several things to successfully meet criteria and be considered compliant. This includes safety program documentation that can depend on the company but will usually include safety programs, policies, and procedures. It can also include other things such as procedures to satisfy client-specific requirements.

If your organization is subject to vetting requirements, either now or potentially in the future, then having an OSH program acceptable to the organization that is vetting your company is important.

Various Other Customer Requirements

Other drivers from your customers can include:

- Requirements of customers when working at their facilities: Your organization may need to supply safety programs or procedures in order to work at customer facilities, as part of site access requirements.

- Requirements of customers when providing construction or field services:
    - Your organization may need to supply safety programs or procedures in order to access projects in the field. This can be a part of an owner's, construction manager's, or general contractor's requirements.
    - If your organization fails to meet customer procedural requirements, or if your organization's safety performance in the field is not acceptable, then your organization's contract could potentially be terminated. This can obviously be a legal matter, but the bottom line is that it can and does happen.
- Requirements of supply chain/purchasers of your products: You may have potential customers that will only do business with your organization if it has certain safety management systems, safety programs or procedures, and environmental management systems in place.
- Project contractual requirements: Many larger construction projects require that contractors and subcontractors submit safety plans, assign designated site safety officers, and otherwise have certain safety processes in place. Many times a process for the completion of job hazard analyses (JHAs), conducting worksite inspections, and holding morning safety talks ("toolbox safety meetings") is required.

Contracts can require things that OSHA does not. It can become a business decision as to whether your company wants to develop documents for projects that are not explicitly required by OSHA or other applicable regulators. Will you not develop a process just because OSHA does not require it?

Safety Related Non-Regulatory Consensus Standards

If your organization is expected by its parent organization, local management, or customers to maintain a certification related to an occupational health and safety management system (OHSMS) or to self-certify to one, then the requirements of the OHSMS become a driver. Examples of OHSMS standards include the ISO 45001: Occupational Health and Safety Management Systems, and ANSI/ASSP Z10.0 of the same name. Occupational health and safety management systems are discussed in more detail in Chapter 8.

As an example, the most common OHSMSs require that an organization have certain safety processes in place, and essentially expect that critical safety procedures will be written.

Teaming Partners

Investors and companies that have teamed with others have a financial interest in the success of their arrangements. Such partners often want to increase the likely success of their ventures by making sure there are no significant downside safety risks. It would not be uncommon for the more financially solvent organization to require that materiel risks be identified and addressed to help minimize risk. Issues that could place an entire organization at risk could stem from the loss of regulatory approvals for key processes, damage to critical machines and processes and infrastructure, and traumatic fatalities or serious injuries. Serious failures could also result in onerous legal agreements that could place a company at a competitive disadvantage and be difficult if not impossible to overcome.

Loss of faith that an organization can continue to win work, obtain permits, and leverage opportunities can result in extra oversight, more difficult financial terms, and loss of key staff. Moreover, appropriately developed and implemented safety programs can help a company to make sure that worker hazards and risks are managed appropriately and that risk to business relationships is minimized.

Other Important Influences

While there are some things that strongly drive or influence the need for creation of OSH programs, there are many things that can influence their creation, effectiveness, and sustainability. Many of these are related to "organizational context."

Organizational Context

This phrase is about the fact that organizations do not exist or operate in a vacuum. There is context to their existence and operation. More specifically, "organizational context" is referring to internal and external forces and factors that impact the environment an organization operates within and is about how those things impact and influence how it runs and how effectively it can meet its safety goals.

The concept itself is important in several management system standards. This includes, for example, standards for quality management (ISO 9001), environmental management (ISO 14001), occupational health and safety management (ISO 45001 and ANSI/ASSP Z10), and others.

Accounting for how organizations operate subject to internal and external factors can be helpful for anyone. Practically speaking, if an organization fails to account for these, it is less likely that their safety programs (and possibly systems) will hold up over time. Figure 2.2 presents a quick review of some other factors and things you may want to think about. Some of these may be more or less relevant to you than others, but the information is shared for your consideration.

These are just a few examples, out of many potential internal and external issues that could affect the success of a safety program:

- Economic conditions can change, driving profits up or down and impacting internal processes. Margins could decrease and result in temptations to

Figure 2.2 Examples of Other Important Influences in Organizational Safety and Health Programs

take unsafe shortcuts or to unsafely bring in new equipment and processes without adequate safety reviews or pick less effective or durable protective clothing.

- The regulatory environment can change and make compliant work either easier or more difficult.
- Technology can improve and make certain approaches obsolete. That can create pressure related to processes and in turn impact the safe performance of work.
- Budget cuts could make it more difficult to maintain and monitor key air monitoring systems and maintenance on important control systems.
- Staffing changes could introduce risk to less well-trained new workers or result in decreased awareness of hazards for employees working prolonged overtime.

The take-home point is not that forces and pressures exist in the workplace from internal and external sources. It's more a matter of having a process to consider how these pressures on the business might impact safety, and then plan for appropriate controls and responses.

Other Interested Parties

Other groups or organizations may have an interest in the success of your safety programs. This could include local business groups or associations. Some associations obtain insurance for a group, like a group of manufacturers. The performance of one organization can have an impact on others. Or an organization could be a critical cog in how several work together. Also, membership in some types of trade associations requires safety plans/programs as prerequisites. This can be true in some surprising affiliations!

Whether all participants in other interested groups realize it or not, they may actually have a vested interest in your organization's safety performance.

Upstream Equipment and Material Suppliers

Visitors to your locations have an interest in being safe when doing so.You should have an interest in helping them keep safe when they are, especially as third parties. Upstream suppliers may gauge the difficulty (if any) and relative safety of getting to your locations and working in your facilities. They may also want to gauge your reliability as a customer. Will you be consistent in your upstream sourcing of their materials, which can in turn impact the efficiency of their production schedules and deliveries? You don't want to be "fired" as a customer!

Third-Party Trades Supporting Your Location

Your organization has an obligation to communicate safety information of certain types to companies providing various services at your location. This includes, for example, energy control programs and hazard communication programs. Communication between organizations can be very important to keep both their and your workers safe. Depending on the nature of the visiting company's work, your organization may have certain other obligations, especially if work of a construction nature is conducted. Depending on contractual and practical relationships, your organization could be considered a "controlling employer" under the OSHA multi-employer workplace citation policy for projects it manages and contractors it brings on site.[7] While that policy is not a law, it provides instructions to OSHA's inspectors regarding what they can or can't cite at multi-employer worksites. Your organization may want to consider how its programs and approaches to safety could be interpreted by OSHA under that policy. Has it taken "reasonable care to prevent and detect violations"?

Local (Non-Business) Community

Your workers may be drawn from nearby communities. Your organization's reputation for safety can influence its ability to hire high-quality talent from the area. Safety programs can help protect your reputation.

Also, will or does your organization provide tours for the local community and other interested parties? If so, it makes sense to establish ground rules for visitors and otherwise keep your visitors safe.

Your property may also be close to local living areas and community spaces. Is there any potential for people, including children, to explore in your area? Will there be vehicles, materials storage, chemical containers, climbing hazards, confined spaces, or other hazards on your site that visitors or trespassers could be harmed by if not controlled and managed properly? Your organization's duties are a legal question and depend upon various factors; the point here is that you should be reasonably aware of potential risks stemming from these sorts of possible hazards and take them into account when developing your safety program.

Distribution Processes, Downstream Channels

If you manufacture and/or distribute materials for sale, your organization may be involved in commercial transportation. Obviously, there can be safety considerations associated with loading trailers, hauling freight, and unloading; or going to locations to pick up equipment and materials and return them to your warehouse, yard, or other locations. These activities can of course present various hazards, either to your workers or non-employees they interface with. Sometimes there can be confusion about relative roles, communication between companies can be less than clear, equipment at different locations could malfunction, and so on. If your organization has an "exposure" to these types of issues, then your safety program should take them into account in an appropriate way.

It could also be the case that after your loads are delivered (by your staff or through contracted companies), people receiving your shipment have questions about things like material handling, stability of load configurations, or the chemical or physical properties of products or materials. While various laws may address various portions of these types of issues, you should determine which issues are relevant to your organization for inclusion in your programs.

Weather/Emergency Management Needs

Even the weather can impact the safety of your workers and influence how work is done. Federal OSHA may or may not publish a heat stress standard, but it has an interest in doing so. Several state programs have issued heat illness prevention standards. While some question the need for a regulatory standard on the topic, uncontrolled heat exposures can have life-threatening implications.

Heat stress prevention programs often include methods to check for and monitor heat illness potential, such as through the use of wet-bulb globe temperature monitoring, the use of heat stress indices, and physiological monitoring. Based on monitoring or the use of indexes, programs can prescribe limitations on duration of work, requirements to take breaks in cooler areas, the provision of fluids to hydrate, and so on. So, the need to protect your workers in environments of possible heat stress can be a driver or influence on your OSH programs.

As we all know, other weather factors can also impact the safety of your workers. This includes high winds, rain, flooding in general, flooding as a risk to outdoor tanks and processes, lightning, and other things. High winds have been known to collapse roofs and walls on top of people, raising questions about design adequacy, emergency evacuation policies and the designation of emergency assembly locations, and other things. Do not disregard or minimize applicable weather-related considerations when developing your programs.

Certifying Bodies

This is related to possible customer or company-driven requirements to become certified to an OHSMS. In the event that an organization chooses to certify to an OHSMS, then the requirements of the certifying organization can influence your organization's safety management processes.

Summary

In summary, organizations need to comply with OSHA's requirements for written programs as they apply, based on the scope and applicability of the regulations. Practically speaking, they should also consider what things they may need or want in their program to help ensure compliance with OSHA's *other* requirements, where written programs or plans are not explicitly required. Understanding the OSHA piece can take some effort. While some applicable standards can be easy to identify, on the other hand, it could take a bit of work to review and understand others.

With a reasonable understanding of the requirements, organizations can usually benefit from developing a bit of a roadmap for needed actions. People who are responsible for a safety program overall do not have to be the primary technical expert for all things "safety" in an organization; it can take a team effort. Depending upon the nature of work, complexity of requirements, size of the organization, and other factors, the organization may need someone or some people to help "translate" requirements for inclusion in an action plan.

However, apart from OSHA, there are many other things to know about safety programs and practical factors to consider. There are many other drivers for having programs, some of which are direct and others more subtle. I would suggest that you consider the broader range of drivers and things that can influence your organization's safety performance. Doing so can help to not only prevent injuries and illnesses in the most effective way, but also to comply with regulatory and other legal requirements. Further, it can really help your team to optimize the return it gets for its efforts. That can include keeping the business running without encountering regulatory or other disruptions or even shut-downs, and leveraging safety processes for business opportunities.

Notes

1 Interagency Agreement between the Mine Safety and Health Administration U.S. Department of Laborand the Occupational Safety and Health Administration U.S. Department of Labor, March 29, 1979.

2 For example, EPA has established some Existing Chemical Exposure Limits (ECELs) under one part of TSCA, and New Chemical Exposure Limits (NCELs) under another part of TSCA.

3 Centers for Disease Control and Prevention, National Institute for Occupational Safety and Health. (2003). *Making Sense of Regulations*. U.S. Department of Health and Human Services. DHHS (NIOSH) Publication Number 2004–101.

4 Occupational Safety and Health Administration. (n.d.). Laws and regulations. United States Department of Labor. https://www.osha.gov/laws-regs.

5 Completion of the OSHA 30 is sometimes required for supervisors on construction and demolition sites by governmental regulators. It is also sometimes built into construction contracts as a requirement for persons filling certain positions.

6 Bureau of Labor Statistics, U.S. Department of Labor. (2025, April 9). The Economics Daily: Median tenure with current employer was 3.5 years in private sector in January 2024. https://www.bls.gov/opub/ted/2025/median-tenure-with-current-employer-was-3–5-years-in-private-sector-in-january-2024.htm.

7 See the OSHA Directive Number CPL 02-00-124, Multi-Employer Citation Policy.

3 Finding Regulatory and Guidance Information

The Good News Is …

> Knowledge is of two kinds: we know a subject ourselves, or we know where we can find information upon it.
>
> Samuel Johnson[1]

As you are developing your organization's written programs, you don't need to know "everything" about OSHA (Occupational Safety and Health Administration) or regurgitate OSHA's requirements into your programs. You need to cover topics required by OSHA, which is often about how your organization will fulfill OSHA requirements. Where detailed and precise procedures might be needed, those can be developed by other technical experts or with appropriate assistance.

Occupational safety is a broad field. No one knows everything. Start by just reviewing fundamental OSHA or State Plan requirements. There is enough information "out there" for you to get your bearings, develop a plan, and move forward.

If you don't already have some knowledge of OSHA requirements and OSH concepts, you can certainly find information about those topics.

There are enough easily digestible things available to get a good start in many areas. Get the big picture, get started, and then you can work on more technical things that might exist. But that can be after you have some ideas for program framework and "where" you might need to dig deeper. Everyone likes a safety leader who has OSH knowledge as well as some understanding of the industry and business they work in, but that doesn't happen overnight.

Some information you will need will be about specific regulatory requirements for which you may need or want written programs or to establish other processes. However, you will also probably want additional information to simply gain a better understanding of safety issues and raise your general awareness and understanding of potential hazards and protections (or "control measures") against those hazards, as well as approaches to manage those hazards.

DOI: 10.4324/9781032704586-3

Types of Helpful Regulatory and Technical Safety Information

You can find good and helpful information from a variety of sources. The trick is not to get pulled into mumbo jumbo that can confuse the issues.

The web pages of regulatory entities can be very informative and to the point. Indeed, starting off on the right web pages can be helpful. This is the case for both national and state sources. There are also other governmental and non-governmental groups that you can obtain information from, as well as non-profit associations and commercial sources.

You could perhaps think of the information as being of three different types. One type is about what an employer needs to implement in writing on a mandatory basis, along with sample programs. Another type of information is about OSH standards that have requirements but don't mandate written programs. A third type is information about safety-related implementation concepts and deeper technical information that one might need when deciding upon how safety processes should run overall, and about other things that can impact program effectiveness.

Consider reading or scanning the information described in the following sections. I mention several current websites. If you conduct a search with the URL I've mentioned and you receive an error message, I'd suggest you search for that same information by title or name. Even though web addresses can change, I don't think the underlying resources will completely go away because of the type of information it is. If at first you don't see what I've described, keep trying – the website administrator may have just moved its location.

From Federal and State Regulators

The national (federal) OSHA website as well as individual State Plans have websites that share useful information.

Federal OSHA

OSHA's website is located at osha.gov. Once on that page, you can find many different kinds of information and resources.

The OSHA regulations and standards can be found there, as can information about OSHA and its programs and activities. It also has information that can be very useful, not only to understand written program requirements for those regulated by it, but also program templates, topical talking points, enforcement information and inspection data, interpretations of standards' requirements (written by OSHA and issued to those who asked), e-tools, information about training requirements, program recommendations, and many other things (e.g., information on injury/illness recordkeeping, and several other deeper-dive topics).

Similarly, the site makes it clear which states have OSHA-approved State Plans, with contact information.

A few good starting points for pulling up helpful information include the following:

- From osha.gov, you can click on the "Standards" menu to see a menu for laws and regulations, Federal Registers, letters of interpretation, training requirements, and more.
- At https://www.osha.gov/laws-regs, you can find the laws and regulations directly. You can select links to see recordkeeping requirements, the general industry and construction standards, and other things.
- See https://www.osha.gov/complianceassistance/quickstarts as a starting point to learn about OSHA's requirements for general industry, construction, and healthcare. This is discussed in more detail in Chapter 6. It gets to the point quickly and provides several other links on one page.
- Sample programs that you can use in support of developing your organization's OSH programs, along with OSHA e-tools to assist with the same, can be found at https://www.osha.gov/complianceassistance/sampleprograms.
- The web address https://www.osha.gov/pls/publications/publications.html takes you to some helpful documents such as the Small Business Safety & Health Handbook and Safety & Health Programs: Recommended Practices and Safety and Health Programs: Recommended Practices in Construction. See the section "All Publications."
- Using the "A to Z Index" located at the top of osha.gov (osha.gov/a-z) is another good way to quickly find information on topics that may affect you. If you look up particular topics, it provides a summary of standards that may apply to those topics, as well as internal directives, letters of interpretation and other things.
- A collection of materials developed under the OSHA Susan Harwood grants program is available for free at https://www.osha.gov/harwoodgrants/grantmaterials/bytopic/. This includes a good number of training materials covering a variety of topics that may be of use to you.

State Plans (States' OSHA Approved Plans)

State Plans sometimes simply adapt the federal OSHA regulations/standards, but others have their own unique requirements and supporting systems. You can review which states have an approved State Plan in place from the osha.gov website. State Plan details can be accessed through those OSHA.gov links.

Regarding written program templates, some State Plans do not provide them at all, yet others provide good starting points. Since there is variation in template content when templates are shared (some are extremely basic while others are detailed and cover more information), you may want to shop around. Of course, your organization needs to ensure that any template it uses (or combination of templates) will meet the requirements for the states it operates in.

If your organization operates in one or more states with an approved State Plan, what it likes or doesn't like as templates can be a matter of culture and opinions (as long as your written program includes the elements the State Plan requires). Note that some states have requirements that require more than federal OSHA and other State Plans.

Examples of State Plan information that may be helpful include, but are not limited to, the following:

- California OSHA (Cal/OSHA) has a wide range of information available on its landing page at https://www.dir.ca.gov/dosh/.
- Cal/OSHA publications including model plans (such as for the state-required Injury and Illness Prevention Program, and others) and safety inspection checklists are available at https://www.dir.ca.gov/dosh/PubOrder.asp.
- Cal/OSHA e-tools are located directly on Cal/OSHA Pubs: https://www.dir.ca.gov/dosh/PubOrder.asp.
- Hawaii has sample programs located at https://labor.hawaii.gov/hiosh/home/for-employers/sample-programs/.
- Idaho has a collection of sample programs, information, and checklists at https://www.boisestate.edu/oshcon/sample-written-programs-and-related-documents/.
- Michigan has fact sheets, forms, sample written programs, and other things at https://www.michigan.gov/leo/bureaus-agencies/miosha/resources/pub-pos-form-media.
- North Carolina provides "OSH Industry Guides" at https://www.labor.nc.gov/safety-and-health/publications.
- At https://lni.wa.gov/safety-health/, Washington state provides links for finding safety rules and creating safety programs, including Accident Prevention Programs, information about safety meeting and safety committee requirements, training requirements and materials, and many other things.

There are other State Plans that provide resources as well; these were just some examples.

From Other Non-regulatory, Non-commercial Entities

There are several other excellent sources of information that are solid. Here are a few examples.

Safety and Health Consultation Programs

You can find on-site safety and health consultation programs based on OSHA requirements in each state, but which are free and distinct from OSHA compliance activities. These are focused on small and medium-size employers. Exactly who runs the program varies between states. A list of the consultation programs and contact information can be found at https://www.osha.gov/consultation/

directory-text. While it is the employer who asks for the consultation, and the consultation program consultants do not issue citations, employers still have an obligation to correct serious and imminent danger hazards identified through the process, within an agreed-upon time frame.

The National Institute for Occupational Safety and Health (NIOSH)

Historically, NIOSH has conducted research that supported standards development by OSHA, worked on research related to the prevention of work-related injuries and illnesses, and developed new knowledge for transference to industry. It is part of the Centers for Disease Control (CDC).

Over the years, much helpful information has been learned by and shared through NIOSH. You may want to scan historical reports, technical information, health hazard evaluations, fatality study data, and more, starting at https://www.cdc.gov/niosh/. Here are a couple of other specific pages you should know about:

- NIOSH Fatality Assessment and Control Evaluation™ (FACE) Reports: https://www.cdc.gov/niosh/face/.
- Workplace Safety and Health Topics: https://cdc.gov/niosh/topics. You could search for workplace design solutions here.
- The OSHA-NIOSH heat stress app is located at https://www.cdc.gov/niosh/heat-stress/communication-resources/app.html.

CPWR

For the construction industry, the CPWR (Center for Construction Research and Training) website at https://www.cpwr.com/ offers a plethora of resources. These range from technical information about the construction industry, to descriptive statistics on injuries and illnesses, to checklists, and templates for developing plans (like exposure control plans to help employers control exposures to respirable crystalline silica [RCS]).

You may want to look at the CPWR's online Construction Chart Book at https://www.cpwr.com/research/data-center/the-construction-chart-book/.

The CPWR site can be a great help in determining what kinds of things you might want to address in your written (and other) programs.

Insurance Companies

Insurance companies and brokers sometimes offer risk management services and may be able to provide you with program development information. When available, it might be possible to obtain a certain amount of program development or informational assistance as part of your policy, or for a fee. I've seen some excellent assistance provided that way, including for topics like electrical safety programs, ergonomics assessments, and machine guarding analysis and recommendations.

Paid Personal and Organizational Memberships

Organizations like the National Safety Council (NSC), the American Society of Safety Professionals (ASSP), and the American Industrial Hygiene Association (AIHA) have a variety of activities and programs that advocate for workplace safety and health and provide various training and networking opportunities. They also publish and/or sell a variety of books and other educational resources that can be helpful.

Other groups that can provide safety related guidance and information include industry-focused membership organizations. These groups can provide safety information focused on a particular industry or technology. This includes, for example, the Association for Advancing Automation (A3) for robotics, and the American Foundry Society (AFS) for foundries and metal casting. There can also be state and regional associations of different types, such as for the Construction Association of Michigan (CAM).

Paid Commercial Memberships, For-Fee

In addition to the above, there are other commercial sources of information that can be helpful for both safety information and program development. For example, there are several excellent products and tools available through Business and Legal Resources (BLR) and J.J. Keller & Associates.

Summary

There are a number of resources available to you for free that provide some form of program development framework. Some of the web pages can be explored for additional context and to learn more around the "periphery" of safety management concepts.

This chapter was not intended, however, to jump into the safety waters too quickly or at any great depth. The following chapters provide much more context, address some practical matters of implementation, and discuss some ways to get the most value from your ultimate program.

Note

1 Boswell, J. (1791). *The Life of Samuel Johnson, LLD.* Henry Baldwin.

4 What Causes Injuries?

Injury Causes Should Influence Your OSH Program Content

> The danger which is least expected soonest comes to us.
>
> Commonly attributed to Voltaire, French dramatist, poet, and reformer born in 1694[1]

If you are developing a company's safety programs, you may well ask yourself what it is that causes occupational injuries and illnesses. A safety program is intended to prevent those, so knowing some information about causes and contributing factors can give you some hints. Knowing that sort of information can clue you in about some things your program should include.

When I refer to "causes," there is the potential to think about direct causes as well as indirect or underlying causes. For that matter, knowing something about types of exposures and processes and types of equipment can give insight here, too. Let's not be too limited in how we think about this. Thinking too narrowly could hide the forest from us, but thinking too broadly could hide individual trees.

It's only natural to have questions about causes. For example, what *hazards* are most likely to be factors in injuries? Should an organization pay the most attention to injuries and near hits[2] that are thought to have the highest likelihood, or just those things suspected of having the highest potential to cause serious injuries or fatalities? Does it make sense to focus the most on unsafe conditions?[3] What about just paying attention to the type of work being done, regardless of data?

I should point out here that the word "hazard" in this chapter is meant to refer to something that can cause injury, illness, or death. This can include conditions, circumstances, or inherent properties.[4]

The answer is that *what to focus on can depend on circumstances and context*: "it depends." That answer reminds me of a particular professor in engineering school who characterized the use of "it depends" as a frequently stated initial answer to engineering questions, and which few people wanted to hear. But it's true. The reality is that there are physical, environmental, and organizational factors that can contribute to an organization's safety and health condition of being a safe

DOI: 10.4324/9781032704586-4

place to work, just as things like forces, pressures, material properties, and energy must be considered in engineering-related questions and answers.

From an OSH (occupational safety and health) standpoint, let's discuss a framework for approaching this, and then consider some injury and illness–related factors.

Learn and Know Your Industry and Processes

I found it interesting when I first got into "safety" that some people seemed to just assume that everyone knew where safety and health hazards existed in different work settings and tasks. And I suppose that if someone has years of industry experience and they have an interest in safety, then they might have a reasonable idea of what hazards can exist in the workplace. Over time they might have seen information from OSHA (Occupational Safety and Health Administration) or companies they've worked for; they might have heard of injuries or received some training. However, for many, that is not the case at all.

Depending on where someone works, the physical environment, and what they have to do, safety is not usually "just a matter of common sense" (as some people like to assert). While in some cases hazards and protection from them can be obvious, in many other cases it is not. This is because hazards are not always intuitively obvious or predictable.

Knowing that not everyone comes from a more experienced background, I'm going to describe more of an iterative learning approach here.

Learning about the processes and tasks your organization has or will have will influence your review of information about what hazards can exist and what types of things can cause injuries. Conversely, once you start reviewing and reading about potential causes, you will probably find yourself circling back to specifications and information about those processes and tasks.

There are many ways to start the process, but a reasonable early step is to identify and learn some macro information about what kinds of processes, tasks, and materials your workers will be working with or expected to perform.

This is not to say that you as the safety program developer need to be the "expert" in all things and hazards. You might be, but you don't need to be. You are helping to establish a framework and direction. By helping the organization to understand some requirements, you can determine or influence direction, while also leveraging the expertise of others who might be more familiar with particular processes and hazards.

Obviously, no one person knows everything about everything. People who are or will be working in areas covered by the OSH program you are developing should be able to and often will want to share information about potential hazards and controls. While some OSH program elements may need to be established (especially initially in the case of new sites developing and hiring rapidly) by existing staff and top leadership, many elements should be developed collaboratively.

One extra point to make here is that, as we explore the processes and tasks that people are assigned to, it can be easy to imagine that work is always performed perfectly, that shortcuts might never be taken, and that errors might never be made. When you ask people what they do, some answers may be very direct and accurate, and others may be based on what the person thinks you want to hear and not match reality. Getting only partial information or "perfect world" information – information that does not always capture how work is "really" done – can leave blind spots. I am suggesting that inquiry should be for more than checking a box and confirming what you think you know. It should be for really identifying potential hazards and things your company might be required to or want to do something about. There is more information about this type of consideration in other sections of the book that talk about performing job hazard analyses (JHAs) (often also referred to as job safety analyses or JSAs), and other types of hazard analyses.

General Injury Cause Considerations and Historical Approaches

After you have an understanding of what kinds of processes, tasks, and materials your organization's workers will be around or can work with, at some point you'll need to delve into specific hazards that can stem from them or work with others who can assist with that.

But hold that thought. I am not suggesting that to write a safety program you need to know design considerations for everything and be an expert in all processes. It's just that at some point an intentional approach to providing a safe workplace can involve evaluating them for hazards. For now, let's just talk more generically about some kinds of things that can cause injuries.

By the way, here is the great secret about what causes injuries: There is no secret. Once again, "it depends." The answer depends upon many factors. There is no universal theorem of injury causation except at perhaps the very highest levels. Also, ideas about what causes injuries have evolved and shifted over time. We don't need to go into that much here, but there are a few concepts worth mentioning that you should probably know about.

One thing to mention is that some people have historically thought of incidents happening because of a "root cause" – as in, "what was the root cause of that?" Some incident investigation forms have engineered or helped create that thinking by asking the same question. The term used to be used to describe a single event or condition that, if it did not exist, would have prevented an incident. Old injury theories supported this. However, it's accepted now that most incidents actually are the result of several different causal factors. It can be a matter of how far and wide one looks, but in systems thinking, processes are related. You don't have to look far, usually. In the case of a saw that had a guard removed and that cut someone's hand, there were probably many other things going on that contributed to the injury, several things that went wrong leading to the injury, and several things that if done differently may have prevented the injury.

A second fundamental thing to mention is that there has been much discussion about causes of injury that revolved around two primary things: unsafe acts and unsafe conditions.

Unsafe acts and unsafe conditions exist and can be related to bad things happening. However, these days, they are often considered to be macro concepts; people think "about" them but also "beyond" them. Thinking about (solely) an unsafe act or unsafe condition will not always lead to discovery of underlying causes and best explanations; there is often more to consider. If those factors are what an organization focuses on, then how will it, in fact, focus on programs to address "unsafe acts?" Knowing of unsafe acts alone does not describe factors influencing those unsafe acts. And, knowing of unsafe conditions alone could be helpful, but does not necessarily convey why those unsafe conditions may have existed or how to prevent them.

A third thing to mention is that, historically, companies have been in a rush after an injury has occurred to have quick answers, and to put together a game plan for obvious direct issues staring at them. "I need to have that injury report completed in less than 24 hours!" It can certainly be good to know quickly what things might still be in place that could hurt other people, but too often in the past the rush was the result of trying to look proactive and tell workers and other interested parties, including higher levels of a company, that things were under control. Company cultures and engrained mindsets often viewed injuries as the result of a single event, or linear events. Organizations would often not evaluate or consider what factors may have existed to drive a worker's decisions to work as they did, nor organizational factors, experiential factors, etc. Did someone lift something in a way that created ergonomic risk, but primarily because of the design of the workstation and standard work method?

Usually there are many potential causes and influences that lead to injuries and illnesses, yet influencing factors and human performance considerations are often still not given attention. There are many articles and studies available through easy web searches about factors that influence, for example, unsafe acts. There are also many (and popular) books on the subject.

This is something that you can and should pay attention to as you frame and implement your program. Keep that in mind as you move forward. For now, let's consider some higher-level types of injury causes and things that you should know about as you build your written and unwritten programs. The following items are shared to give you items to think about.[5]

Note that injury causes of highest importance to your industry may be significantly different to those of importance and frequency from other industries. My intent here is to share some information that applies broadly. It is not possible to list all sources for all industries and all situations, but this should give you some ideas.

Some Causes of Injuries, in No Particular Order

1. Exposures to unsafe *machine operating conditions and work tasks.*

This can include injury from getting pinched by, crushed by, or pulled into hazardous machine parts, or the release of hazardous stored energy (like from capacitance discharge), the release of hazardous chemicals or radiation, and noise, sharp edges, and hot surfaces. Unguarded hazardous motion and unexpected motion are frequent culprits.

OSHA has various standards that address the basic issue of machine operating conditions, including those for machine guarding, energy control, electrical and other standards. (Energy control is discussed more in item no. 6 below.)

People are sometimes amazed to hear this, but when machines or equipment are sold in the United States, there is no *OSHA* certification or approval process for it. Some people have asked me over time, "can't I just presume that if the product is sold on the market, it is going to meet OSHA requirements for our employees?" The answer is, "no you can't." OSHA has requirements that employers provide a safe workplace, but OSHA is not in the business of certifying or approving equipment, and safety needs to be ensured by the employer. (Yes, manufacturers can have product liabilities, but that does not stem from OSHA and is beyond the scope of this book.) Some manufacturers build things with varying levels of safeguards. They sometimes count on purchasers to let them know if they want more safeguards, or to otherwise use equipment safely. Unfortunately, I've heard from too many manufacturers, "if you want [name the reasonable control you'd like to see], the company can upgrade but that will cost more."

You probably see or can imagine that many actions can be taken to prevent injuries from the types of things just discussed. You may not have the background yet to know how to specify or roll out all possible controls. However, for later possible use, here are some examples:

- The issuance of design specifications, as well as conducting a pre-purchase design review (for guarding, machine safety controls, compliance with ergonomic design guidelines, and other things)
- Team-based safety reviews during installation and process trials
- Initial and periodic equipment inspections looking for critical installation/equipment/integration deviation
- Internal change management requirements that safety reviews be conducted if changes could result in safety risks
- Worker qualification and training
- Post-injury and post-incident reviews
- Energy-control/lock-tag-try programs
- Consideration for practical machine access
- Many other things

2. Exposure to unsafe *walking and working surfaces*.

Many have heard of the issue of "slips, trips, and falls." However, not everyone appreciates how very real and important the prevention of this type of issue is, including falls from height and even falls that happen on the same level that someone is working on.

Uneven walking surfaces, inconsistent construction of stairs (riser height and tread depth), the unsafe use of ladders, slick surfaces, and performing work tasks at locations with fall potential but that have no fall prevention or protection can all be things to look for and address. I'm aware of people who have broken bones just from slipping on oily floors, or by falling off foundry melt decks that had no fall protection at the edges of elevated walking surfaces, or falling through holes in construction environments. I've also read many fatality reports and investigations related to people falling from roofs, scaffolds, through skylights, and even just falling and hitting their head on a same-level process pipe.

How can organizations prevent these sorts of issues?

First, you need to know where your organization's slip, trip, and fall hazards exist or might exist, or could be created (e.g., guardrail systems that could be removed but not replaced).

Second, know that, based on that information, OSHA and OSHA-approved state programs ("State Plans") have a variety of regulations that are intended to help prevent these issues from happening or to otherwise protect from injury. Organizations are required to comply with these.

For employers regulated by the OSHA general industry regulations, OSHA describes walking and working surface requirements (including fall protection) in 29 CFR 1910 Subpart D and Subpart R (fall protection applicable to work related to electric power generation, transmission, and distribution). Height requirements for fall protection can vary based on industry application. OSHA lists the Construction Industry regulations at various locations in 29 CFR 1926, and requirements can vary based on the type of work or surface one is working on (such as for steel erection, scaffolding, and leading edge deck work). State OSHA programs often defer to the federal requirements, but state programs do sometimes have more stringent requirements. Recall that you can find a list of the federal OSHA regulations at https://www.osha.gov/laws-regs/regulations/standardnumber/1910. Be sure to read them.

Based on applicable OSHA regulations and your company's type of work, your organization may be subject to specific fall protection–related requirements. However, what OSHA requires to be written may be less than what your company feels it should describe in writing. There can be cases where your organization would not be required by OSHA to describe certain processes in writing, and yet since OSHA still expects the organization to "do" certain things, it might choose to place expectations and requirements in writing anyway. That's a matter for your organization to determine.

While requirements can vary based on industry and type of work, it's still the case that organizations should have processes in place to comply with the relevant OSHA requirements. The organization may choose to require certain things to help ensure that it is compliant, or to require actions or protections that go above and beyond what OSHA might require.[6]

OSHA requires organizations to conduct assessments to determine locations and types of work that do or could be subject to slip/trip/fall potential and corresponding OSHA requirements. Apart from that, organizations often also require prompt spill cleanup for slip prevention and exposure protection reasons (apart from potential environmental regulatory requirements). They may place special emphasis on the effective containment of oil from machining processes for slip prevention. They could conduct inspections to ensure that structures remain as designed for fall prevention. They sometimes modify walking surfaces to make them less slick to walk on and could require special non-slip shoe soles. Depending on the nature of the process and task, they might even be able to redesign things in order to avoid needing to work at elevated heights (if they did not have the chance to address that previously).

At this point, I'm sure you get the gist – when hazards are identified, OSHA regulations may apply and mandate certain actions. But even when there is not a specific OSHA regulation that applies, the organization may still need to, and/or want to, require other protections for injury prevention.

How will you know when the organization should actually do things that are not explicitly required by OSHA, or not required by OSHA but that could help protect your workers? The answer to that is tied to OSHA General Duty Clause requirements, and hazard identification and risk assessment processes – covered elsewhere.

3. The performance of work tasks that overly *stress* the body – you should not disregard this.

Some work tasks may require awkward postures, forceful exertions, or high repetition. Those may create a risk of injury to muscles, tendons, nerves, soft tissue, or cause other damage.

Before processes are designed and installed at a work location, designing processes to meet generally accepted ergonomic design principles can help. Once processes are designed and installed, the organization may want to conduct assessments to determine if there is in fact ergonomic risk that could cause injuries. The evaluation may indicate the need to identify and evaluate improvement options.

From a program standpoint, what your organization does regarding ergonomic risk factors can depend on what is being planned versus what already exists, injuries that might be occurring, and levels of risk. Know, however, that if you have only newly started certain processes or tasks, just because the organization is not receiving complaints or actual reports of injury, it does not mean that significant risks do not exist. It can take time for certain injuries to occur.

When they do occur, they can occur suddenly. Not only can it be quite a bad experience for the employee, but it can also be very costly and impactful to the organization.

The bottom line is that we know work tasks can pose ergonomic hazards and cause injuries. The question becomes, what does the organization need or want to do about that? At this time, federal OSHA does not require that employers have a written "ergonomics program," although some state programs do in various contexts. It is also the case that some employers have been issued citations for ergonomics issues under the General Duty Clause.

Regardless of OSHA requirements, there is no question that ergonomics-related injuries represent a large share of the types of injuries that occur, and that the costs are tremendous.

4. Exposure to *unsafe levels of harmful chemicals*. Many people know that exposure to *harmful chemicals* in the air (or through the skin or by ingestion) can have a range of toxic effects that could impact exposed workers immediately, over short time frames, or over long time frames.

OSHA publishes legal limits for exposures to a range of air contaminants. Some limits are contained in a table and are concentrations in air not to be exceeded over a specified period. However, OSHA also publishes some substance-specific standards, such as for lead, which not only specify air limits but also many other requirements. There are different types of concentration limits. Some are referred to as "Permissible Exposure Limits" and "Action Levels." There are also other types.

Not everyone knows what to do about those "funny smells" in the workplace, or whether or not those short-term but maybe chemical-intensive things they do is really a problem (or not). People should consult applicable safety data sheets (SDSs), which are required for many materials (related OSHA Hazard Communication program requirements are located at 29 CFR 1910.1200 and 29 CFR 1926.59). Beyond that, conducting exposure assessments may be either mandated by OSHA or warranted based on exposure potential, but that can be a technical and complicated process. An evaluation by a competent industrial hygienist/occupational hygienist can be helpful to assess if exposures are within regulatory limits or guidelines and to make exposure control recommendations. The bottom line is that it is up to employers to determine if exposures are within legal limits, and some organizations choose to be even more protective.

There are several different commercially available software packages that can help manage chemical information and exposure monitoring results.

Note that while exposures to chemicals could happen anywhere in a workplace, certain types of what OSHA calls "confined spaces" (the OSHA definition has a specific meaning) can have special air quality considerations. For example, when chemical liquids, harmful dust, or gases are in them, they can (depending on the material) present extra toxicity, oxygen displacement, and flammability risks. Some materials can displace oxygen or reduce oxygen levels and create atmospheres that are dangerous or immediately dangerous to life and/or health (IDLH) (see 29 CFR 1910.146 and 29 CFR Subpart AA).

Organizations should have processes appropriate to meet OSHA's hazardous materials–related exposure limits and requirements.

5. Exposure to *hazardous noise* can cause noise-induced hearing loss. Many people know that prolonged exposure to hazardous noise can cause injury resulting in hearing loss.

There are different kinds of hearing loss. Not everyone knows that permanent *noise-induced* hearing loss is irreversible, however, because it damages the hair cells located in the inner ear. People should not think they can "tough out" the noise during their working years and then just wear hearing aids if needed after they retire. While people may use hearing aids to try to hear better, their original level of hearing will never return.

Noise levels that exceed certain eight-hour time-weighted average limits require that employers implement hearing conservation programs (HCPs), including audiometric tests, training, and other things. You won't know for sure if you exceed the levels of noise requiring an HCP or mandatory hearing protection unless you conduct valid noise testing in the workplace. Consider conducting sound level checks and noise dosimetry (which integrates the sound pressure levels over time to create a time-weighted average) to determine your organization's status.

As a practical matter, there are some sound level meter apps for phones that in some cases can be used as a tool. However, they are not all of the same quality or accuracy. The National Institute for Occupational Safety and Health (NIOSH) developed a sound level meter application (app) for iPhones (running iOS) (under memorandum of understanding with EA Lab), which met Type 2 requirements of the IEC 61672:3 standard when calibrated with an external microphone. Phone apps for use with Android phones did not meet that standard. While some apps may be helpful, they can also underestimate noise levels. Make sure that testing to determine your compliance with OSHA limits is done using equipment that fully meets OSHA requirements.

6. Exposure to sources of *uncontrolled hazardous energy*. People get killed every year due to the unexpected release of hazardous energy coupled with the exposure of a worker or workers to that energy.

Hazardous energy can exist in machines and equipment, but in other locations as well. It can take the form of hydraulic, pneumatic, chemical, thermal, electrical, and others.

Energy is needed to make things "happen" in industry, and there can be "good" energy as well as "bad" energy that can harm. The idea for you in your safety program is to assess and know where various types of energy are, to ensure the right precautions and training are determined, and to have the programs, processes, and procedures required in your organization to help keep people safe.

OSHA has several standards related to this, but depending on how you define "hazardous energy," you can find requirements in different places. For example, one standard is at 29 CFR 1910.147 – the control of hazardous energy (lockout/tagout) – which covers "the servicing and maintenance of machines and equipment." But there are others that are related, like 29 CFR 1910 Subpart S – Electrical; and specifically 29 CFR 1910.333 – selection and use of work practices – as well as many others to include for the construction industry.

7. *Unrecognized hazards.* Workers cannot perfectly predict hazards and also don't always realize when they are shifting away from what they are supposed to do (although sometimes they do; see no. 9 below). This is not something tracked in statistics kept by the Bureau of Labor Statistics (BLS), which focuses on more direct injury and illness information, such as injury events and sources of injury. However, many OSHA citations reference an employer's failure to recognize hazards. And I've personally seen employers' and employees' failure to recognize hazards as an underlying factor for injuries in companies I've worked with over time.

Note that hazards might be unrecognized by employers, and hazards might be unrecognized by employees, or both. Or they might be recognized but breezed past and not thought to be significant enough to do something about. Those are different things and can have different causes and implications. When the OSH Act in Section 5(a)(1) talks about furnishing a workplace free from recognized hazards, it is talking about the duties of the employer.

There is no across-the-board requirement in federal OSHA's regulations to conduct a job hazard analysis (JHA), as so named and described in OSHA guidance publication OSHA 3071, Job Hazard Analysis, even though it is common practice across industries, and for many employers, to complete them. Some state-run programs have requirements that mandate hazard assessment, evaluation, or controls, although not explicitly, in a JHA format.

However, there are various (Federal) OSHA regulations pertaining to specific topics that require some form of hazard identification, inspections, and assessments to determine appropriate actions. The completion of JHAs is just one tool available to employers to help identify and control task-based hazards that can cause injuries and illnesses. Many other and more advanced tools are available for hazard identification and risk analysis. Some focus on task-based risk assessments, while others are focused more on process hazards but may include consideration of potential sources of error from the accomplishment of particular tasks.

The point here for you is that many good hazard control practices begin with recognizing where hazards exist. What processes will you build into your safety program to do this?

8. *Rapidly changing conditions.* For some types of work more so than others, conditions can change rapidly and move from being relatively safe to unsafe, with no or limited detection until it is too late.

This can be related to the "unrecognized hazards" mentioned in no. 7 above, but no. 8 is more about changing conditions and how people respond to them. Some issues in potentially rapidly changing environments can include the potential to see changes but accept increasing risk to obtain higher rewards, a failure to understand gaps in safety that can be created during the changes, and the potential to perform work more or less on autopilot unless something really stands out to grab the attention.

Have you ever been engaged in a task without giving it much thought, but then concluded that what you were doing was not ideal from a safety standpoint? Have you ever been faced with a decision about either finishing work or going back to square one?

I've known cases where workers were in situations that were not as they originally anticipated, with inadequate advance planning and flexibility, and the situation resulted in fatalities. Safety programs and approaches should account for dynamic situations and incorporate methods to help counteract or control this potential situation. Especially for this sort of work scenario, consider building processes that take human performance into account. This can include that humans routinely make errors, and that systems influence behavior.[7]

9. High levels of *risk tolerance and risk-taking*. I think there are a few different scenarios to consider.

One is that there are some people who are natural risk-takers and thrive on being able to take risks. Another is the situation where people intentionally take risks and believe they can safety do so, or there are other "drivers" or factors that can make that "OK" within the local culture.

These can be deeper topics and relate to anything from the sheer joy of breaking rules and taking risks (that "some" might have), to other cultural and systems-based reasons. The answers to how to best avoid injury as the result of these could vary.

Managing these can be a matter of *how* rules and safety processes get applied, as much or more so than what the procedures on paper are. Some situations might need to be coordinated through a human resources department. However, managing this should also include considering how your organization thinks about human performance and causes of error. When developing your program, as a practical matter think about your organizational structure and how workers will or may interact with processes and tasks from a human nature standpoint. A couple of commonly emphasized points would be to make sure it is not the organization's systems that are driving risk-taking behavior. Ask "why" things are taking place, from the standpoint that at the time a worker makes a decision, there is some sort of rationale or basis. Responses from leadership should be well thought out, deliberate, and result in learning from what goes right as well as what goes wrong.

You can keep these considerations in mind when conducting inspections or hazard analyses, when conducting a post-incident review, and determining what the organization establishes or models for its other processes. Think in advance about the framework the organization will use for evaluating different types of risk-taking behaviors.

10. *Human errors*. Of course, people make errors. Sometimes that can result in injuries. No great surprise, right? But there is something to point out here.

It used to be the case that when incidents occurred, much of an organization's response could be not only on whether there was some sort of mechanical or physical failure in something, but also on whether someone did something wrong that caused that failure. In and of itself, that can be a valid consideration – if someone was operating something out of known tolerance, for example. However, sometimes the intent was to blame the injured person as opposed to learning from what happened.

There seems to be more recognition that error is normal, and that not all errors result in injuries or incidents. This is incorporated into safety management styles and systems. More organizations consider and account for likely errors when designing processes.

Processes should not be designed to expect human perfection and inexhaustibility. This can be a big concept and have many tangents of discussion, including for equipment design.[8] While not addressing legal concepts related to product liability, the idea is that while some tasks can demand high attention, accuracy, precision, and repeatability, ultimately we make mistakes. We can attempt to design out mistakes and the impact of making those mistakes, in the way we design equipment and safeguard that equipment. However, we can also take that into account when we think about the human behavior and error potential in safety program management.

Let me make an analogy to the process of jumping rope. Okay, this is not a perfect analogy, but please bear with me; I think it's worth making this point.

Imagine an athlete jumping rope to achieve fitness and increase coordination. When they are first learning, they may not complete too many jumps before stumbling on the ground or tripping on the rope. They have the best of intentions, but they are only so good. Missing a jump can be considered similar to making a workplace error. As they practice over time, they might be able to jump the rope for longer periods of time, at a faster rate, and maybe even make trick jumps. But there still is a point in time where the person loses some focus, or loses energy, or they lose their timing and essentially make an error. They can't jump forever, with perfect repeatability, or ultimately without some form of error.

When you develop your safety program, you can do so with the mindset of considering and adjusting for human performance factors. You can also design and manage the program to focus on learning instead of blaming.

11. *Conflicting workplace demands*. Granted, this is a bit more removed than a direct cause of injury. Even so, it is something that can contribute to injury.

Many people have worked in situations where there were conflicting demands placed on them and roles were unclear. That kind of issue can cause stress and make truthful and effective communication less likely. People can question what they are "really" supposed to do, where priorities lie, and what will be rewarded.

That in turn can result in less focus on the job, a breakdown in order, and even create situations that others would not predict or expect.

You may have heard about the warehouse worker who was told not to speed, yet it would be impossible to meet truck loading expectations and shipment goals unless forklifts were driven too quickly, barring divine intervention. Or the maintenance mechanic who was pressured to fix a production machine very quickly, and did so quickly enough that lockout procedures could not possibly have been followed, and was commended by the supervisor with the tacit understanding that shortcuts were taken. Those kinds of situations promote taking inappropriate risks and a workplace culture that promotes production over safety. That ultimately leads to problems. It could be a sign of misaligned messaging from senior management, "doubt" that safety is a value, and discrepancies in what activities and metrics get rewarded.

Preventing this kind of issue can take significant intentional effort from multiple directions and levels of management, and especially regarding the creation and maintenance of a culture that supports safety. It can help to use a systems approach to management. Key areas of attention can include ensuring leadership direction and support for safety, worker participation, the establishment of aligned goals and objectives, the integration of safety processes with those of the balance of the organization, audits and self-inspections, and clear management dialogue.

How many of these things are directly required by OSHA? That's a rhetorical question.

Note that there are increasing discussions about workplace stress and *psychosocial risk factors* in the workplace. The ANSI/ASSP/ISO 45003-2021 standard[9] (a nonmandatory document that provides readers with guidance about managing psychosocial risk in organizations using the ISO 45001 management system) talks about how these sorts of hazards, in the workplace, can result in injury and ill health.

What are psychosocial hazards? NIOSH defined them as factors that cause stress, strain, or interpersonal problems for the worker.[10] They are psychological and/or social factors that can in turn cause worker psychological or physical harm. They can stem from a variety of sources. The standard discusses those that may be related to work organization, social factors at work, and the work environment, equipment, and hazardous tasks. These hazards can impact a workplace's safety, even though it might be difficult to summarize neatly. Organizations can choose to consider how psychosocial risk factors can influence worker safety and manage accordingly.

More General Causes but Extracted from National Data Sources

Here are some more causes of injury as extracted from a few different national data sources. Note that there is some potential for confusion of terms here, due to the fact that different organizations collect data for different purposes and scopes, and our own preexisting ideas of what the words mean.

For example, there can be some confusion when we think of "sources" of injury as opposed to "types" of injury. Organizations that publish national or regional data have to have definitions and rules for what injuries to count within those definitions. Performing searches for causes of injury as published by, for example, OSHA or (formerly) NIOSH tends to yield high-level statistical data sorted by "leading causes," industry and other factors, but the "causes" are quite broad. For example, "causes" could be "motor vehicle related incident," "machine related injury," and "fall," which can have many different underlying details. The word "cause" in this context does not go very deep.

The following eight items are a consolidation of higher-level causes of fatalities and nonfatal injuries. These items are similar to the earlier list but extracted from national data based on frequency. They are not listed in a particular order. The information can be more or less important based on industry. The bottom line is that if the shoe might fit you, wear it by looking at how it may apply to your organization, and planned or actual operations.

1. Caught in or struck by equipment/machinery
2. Falls (including slips and trips)
3. Exposure to harmful environments
4. Overexertion (ergonomic related)
5. Transportation incidents (motor vehicle crashes)
6. Assaults and violent acts
7. Machinery incidents
8. Industrial vehicle incidents (including forklifts/powered industrial vehicles)

OSHA has a list of "Fatal Four" (aka "Focus Four") causes of fatalities in construction. The list has existed for many years. These four are not the only causes, but they represent a significant percentage of them (about 65% based on data published by the Center for Construction Research and Training [CPWR]). They are highlighted in training courses provided as part of OSHA Outreach Training programs.

1. Falls
2. Struck by object
3. Electrocutions
4. Caught in/between

Each of those four categories has subcategories. For example, while falls are listed as no. 1, there are many kinds of things that people fall from (leading edges, scaffolds, same-level, and other things). Also, of course, while those four categories describe frequent causes for fatalities in construction, there are other causes of both fatal and nonfatal injuries that programs should address in some way, even though they aren't in that focus list. This includes highway work zone and site traffic control incidents, heavy equipment runovers, forklift and mobile elevated work platform related incidents, crane incidents (including collapse, dropped loads, and electrical contact), trench collapses, and others. The causes of fatalities can be different than the causes of nonfatal injuries.

You can learn more from the OSHA website (osha.gov). The website cpwr.com also publishes chartbooks (look for the latest version of "The Construction Chartbook") that share information about the construction industry as a whole and in respect of various injury statistics.

For an organization to know these categories can be helpful. However, it is still the case that the nature of work performed, equipment used, and other things create specific and unique hazards that the organization might not think about unless it conducts some form of hazard analysis. More on that later.

OSHA Citation Data

OSHA publishes some of its inspection data from federally regulated states on its website (osha.gov). This includes a summary of citations issued within a given fiscal year (October 1–September 30) per regulation number (meaning regulations believed by OSHA to have been violated by employers) along with the dollar value of the corresponding penalties. It provides the public with an ability to sort the citation data by industry type (using the North American Industrial Classification System [NAICS] codes) or standard cited. That could be useful to you.

As we know, OSHA regulations are intended to help prevent injuries. OSHA shares its inspection data with the intent to help raise awareness with employers and then, in turn, avoid injuries.

This annual list as published is aggregated and spans industries. The reader will see both "General Industry" and "Construction Industry" standards in the list. As you develop your safety program, you can look up the citation data that is specific to your industry. The results for your industry can look quite different than the annual OSHA aggregated summary; it can be quite revealing to pull up data for your organization's particular NAICS code. By the way, the data is broken down to the 5-digit NAICS code level, which helps bring the data closer to what your organization actually "does."

The standard top ten list is almost exactly the same, year after year. A citation type might be listed higher or lower in one year versus another, but we tend to see the same categories. Should you care that much about whether a citation was listed as no. 3 versus no. 4, and so on? Probably not. The important thing is to see the categories that are getting cited. If you search the OSHA data, you can also see how many times specific standards were cited, and the correlated penalty amounts.

The most frequently cited standards in fiscal year 2024[11] were:

1. Fall protection, Construction (29 CFR 1926.501)
2. Hazard Communication, General Industry (29 CFR 1910.1200)
3. Ladders, Construction (29 CFR 1926.1053)
4. Respiratory Protection, General Industry (29 CFR 1910.134)
5. Control of Hazardous Energy (lockout/tagout), General Industry (29 CFR 1910.147)
6. Powered Industrial Trucks, General Industry (29 CFR 1910.178)
7. Fall Protection Training, Construction (29 CFR 1926.503)
8. Scaffolding, Construction (29 CFR 1926.451)
9. Eye and Face Protection, Construction (29 CFR 1926.102)
10. Machinery and Machine Guarding, General Industry (29 CFR 1910.212)

So what might you be able to do with this information?

It can make sense to review citations for your industry to help it both avoid potential citations as well as to address potential exposures that could result in injury.

With a bit of an idea of what OSHA cites for your industry or types of processes – perhaps some information based on research from within your industry, perhaps from some free or for-fee checklists that you can review online, perhaps some injury information you've read about or can obtain from similar sister facilities – you can start to develop an idea of specific things that could go wrong and hurt people.

There are some excellent and very detailed books that discuss potential hazards from various facilities, industries, business processes, and exposures, as well as approaches to control, that are far beyond the scope of this book.[12] It is suggested that you review other and more detailed safety texts to gain ideas and possibly identify hazards that you may not have thought about.

Workers' Compensation Data

If your organization has been established for a while, it should have historical injury data available to appropriate staff. Workers' compensation (WC) data could be helpful in some ways and less helpful in other ways. If you obtain workers' compensation loss runs and special reports from your insurance carrier, there are some conclusions that can and can't be drawn, or should be made very carefully.

Sometimes excellent information can be available. However, the data can be at a very high level and is only as good as what was available for entry into the system. Reports can provide injury categories and external "causes" of injury, but the cause categories can be very broad. In my personal experience, it was often too vague to be used by itself and needed to be accompanied by secondary information and extra research. Information necessary from a safety manager's perspective was not always available.

Injury "cause" codes entered into the insurance-based injury reports that I reviewed were based on standard insurance industry coding categories. The data provided categories of information that were helpful to some extent. However, they were not intended to provide a review of underlying incident causes, nor did they apply a systems approach.

You probably won't find out, for example, if an injury/illness stemmed from a collection of things or combination of factors that went catawampus and permitted a particular unsafe environment to exist or a particular decision or error to be made. Be careful and know that if you only consider what someone was doing immediately before they are injured, you might be missing out on seeing a huge part of the landscape.

For example, for a particular injury case, a report could list a cause code indicating that the injury resulted from being caught/crushed in/between objects. That could be helpful, but extra information might be needed, such as through a narrative or potentially a separate injury report. By the time information is available in a database, it can be difficult to have the time to dig further. Also, injury data might not always be coded consistently, as a factor of communication between those reporting the information, treating providers, and insurance company staff.

That said, workers' compensation–related reports certainly *can* be helpful in various ways. For example, I recall that for one organization, the largest category of injury-related costs stemmed from ergonomically related injuries – musculoskeletal disorders (MSDs) and repetitive strain injuries (RSIs). It was good to know that. There was enough information to dig further and ensure ergonomics assessments were underway. We didn't have all the answers, but we knew some areas to emphasize.

Reverse Engineering Regulatory Requirements

In addition to the above, organizations can learn from applicable OSHA regulations. You might wonder what the point would be of thinking further about injury causes if there is an applicable regulation, since we have been talking about the use of injury causes to in fact determine safety program development needs. In Chapter 2, we discussed some of what needs to be developed and incorporated in written programs based on industry type, work performed, chemicals used, and other things. The standards were developed initially based on someone having been injured (at some point) from the lack of protections that those written programs can provide.

The answer is that there is a gap between what OSHA can regulate and what hazards might remain.

By knowing about and scanning regulations that might be or are applicable to you, you may be able to gain a broader sense of the types of issues the regulations are trying to address. A review of the regulations could give you ideas you hadn't thought of; you may wind up incorporating things into your written programs based on those ideas.

Federal OSHA and state-run OSHA programs can't specify everything. It is impossible from a practical and governmental standpoint for regulators to write standards for every possible cause of harm and risk in the workplace. Industries and technologies change; exact work scenarios differ. It would not be feasible or possible for OSHA to predict all things, nor to address every variation of employers' circumstances.

Where OSHA has not promulgated specific hazard or industry-focused regulations through rulemaking, an employer is still obligated to "furnish to each of his employees employment and a place of employment which are free from recognized hazards that are causing or are likely to cause death or serious physical harm to his employees," as you know, in the General Duty Clause.

The bottom line is that just because an employer may comply with specific OSHA-promulgated standards contained in 29 CFR Part 1910 or Part 1926, it does not mean that there may not be other hazards from the workplace or from conducting work that an employer needs to address in some way.

A critical question for your organization is whether it wants to set its safety bar to merely comply with OSHA's requirements (specified legal obligation) or to more fully define how it wants "safety" to work and to identify and take advantage of safety-related opportunities.

For self-preservation, organizations should strive not only to comply with the subject-specific standards of OSHA (covered earlier in this book) but also to address other types of OSH risk not explicitly regulated. Worker safety risks are referred to as "hazard risk" in some risk management texts and can be not only impactful to workers but also related to other forms of risk that are important to the organization. It is one element of "enterprise risk."[13, 14] Do not fall into the trap of simply accepting OSH risk as an unavoidable "cost of doing business."

Other Possible Sources of Information

As you investigate injury causes that may be appropriate to build into your program in one way or another, and if you are part of a larger organization, you may already have injury data for your types of processes, but from sister plants, available to you.

Information about frequent causes of injury in your industry might also be available from industry and trade association magazines and other communication. There are also free magazines available in paper format or via the internet that provide articles about various types of hazards, preventive measures, program considerations and other controls.

Information about potential exposures to hazardous materials that can result from work locations and tasks in particular market segments has been published by the American Industrial Hygiene Association (AIHA).[15]

You could also consider joining and participating in organizations and associations centered around occupational safety and health. That includes state-based organizations and safety conferences, as well as, for example, the AIHA (as mentioned above), the National Safety Conference (NSC), the American Society of Safety Professionals (ASSP), and others.

Artificial Intelligence

Yes, artificial intelligence is being applied in the occupational health and safety world, like so many others! The best is probably yet to come, but there are already AI tools that can be explored for use in program development and beyond. These tools include generative AI, computer vision, predictive analytics, and natural language processing. Tools specific to environmental, health and safety (EHS) are available for both manufacturing and construction.

AI tools can do things like answer safety questions about specific topics, and in the context of this chapter, they can provide information about hazards, injuries, OSHA requirements, and other potential program development needs. We can also use AI to generate safety manuals, scan images for safety risks, detect serious injury and fatality (SIF) precursors, use in video surveillance for real-time notification of risks (e.g., smart cameras, smart hardhats, smart boots), analyze video for ergonomic risk factors, provide data-driven predictive analysis, make OSH recommendations, review uploaded programs and procedures for gaps, analyze reports for trends, and other things.

AI tools still need to be verified for accuracy and completeness. In my experience, as of this point in time, AI-generated information can have a tendency to pull from invalid original sources of data and make conclusions that are not always correct. However, they can be a phenomenal help in conducting research and sorting through large amounts of information, as well as in providing for a higher quality of completion of forms and reports and accomplishing various other management tasks.

Various systems are available and can be applied – from an "entry" to "enterprise" level. Obviously, there can be huge differences in price and what any one product can provide. Some could be cost-prohibitive to a smaller organization, but on the other hand, some can be of significant value for a reasonable cost.

Hazard and Risk Assessment (JHAs and Other Tools)

The information about historical causes and sources of potential injury and OSHA-related information that we just reviewed above can give you some things to think about as you plan out your OSH program.

However, your organization could have other hazards not previously mentioned or that were not yet identified to have caused injury, which should be considered in the management of safety. What we just reviewed can make you aware of some things that could apply to you and maybe have caused injury in the past. But there might be some other potential causes of injury that you still don't have on your organizational radar.

While looking at old injury data and potential causes of injury can give you some clues about things to look for, there can be other potential hazards that you might not have identified from that more general information, unless you have specific processes and work tasks in mind.

As you know, there can be hazards in a workplace for which there is not a specific OSHA hazard-specific standard. We discussed how even where specific standards do not exist, OSHA can sometimes cite employers for violating the General Duty Clause of the Occupational Safety and Health Act, if certain conditions are met (see Chapter 2).

That being the case, it would make sense for employers to have processes to know where else those hazards do or could exist, to be able to protect workers from them – right?

We touched on the potential to conduct JHAs/JSAs in the section above regarding unrecognized hazards. The completion of JHAs can be one helpful process to evaluate job tasks to identify and prevent or control hazards that could result in injuries. That can include hazards for which there are regulations, and hazards for which there is no explicit (topic-specific) regulation. So, let's cover a few more details about JHAs.

There are several guides and templates for conducting JHAs that can be found. That includes OSHA Publication 3071 (referenced above) and many other publications. The JHA development process described in Publication 3071 is presented in a simple form. It includes information on potential hazards and controls.

Many organizations mandate an internal process to conduct "hazard identification and analysis/assessment" in some fashion. The completion of JHAs can both help identify tasks with hazards needing controls and be of value in the development of safety programs (in early planning processes), as well as be "built into" an organization's safety management program for ongoing and periodic use.

A simple JHA often consists of a one-page template with boxes to fill in by a person or team. With limited space on a given page, the work tasks selected for analysis will obviously be limited. What are the boxes, and what information gets filled out? Elements of JHAs often include the following:

- Identity or description of work process
- High-level description of task steps needed
- Identification of the hazards that could be of concern during the accomplishment of those task steps
- Identification of needed controls

The process for and format of the hazard identification tool that an organization uses (e.g., JHA) can depend upon industry, and whether processes are somewhat routine and standardized, or if they are more project-based and variable. It can also depend on whether the organization wants to include information about estimated levels of risk. More on that later.

As an example, a manufacturing organization may develop JHAs for employee tasks at production processes that are expected to be stable and repetitive. The JHAs in that case might be prepared initially, then reviewed periodically thereafter for accuracy, but perhaps only periodically unless there is a process change. On the other hand, work of a more non-routine nature, and especially those involving field projects, can require a different and more dynamic approach. An excellent summary of risk assessment types, including project-oriented considerations, can be found in Popov et al.[16]

End-use of JHAs could vary based on circumstances, but in my opinion the controls specified on the form should be the controls actually required by the organization or supervisor to perform in order to do the work, unless a few acceptable options are clearly identified as such. (Some organizations slip into a mode of listing what someone "maybe" could do, but then workers tend to view the JHA as optional and disregard it.)

My point in bringing this up is that the JHA process can identify controls that should be built into the organization's safety program. For example, required controls could be to follow particular procedures, utilize particular work methods with specific tools and types of PPE (personal protective equipment), and provide training. The safety program could talk about preventive maintenance required to inspect and maintain those controls.

Some organizations create JHA templates that have "memory joggers" built in that can aid in completion of the analysis. This can be particularly helpful in certain industries.

More complicated JHAs are sometimes structured by organizations to take a few pages and can capture more information than the basic JHA. This is especially the case if the JHA is modified in order to require estimated risk levels. In those cases, the form might be called a job risk analysis (JRA) instead of a JHA. Many different approaches are possible, and the best choices will depend on the organization, other processes that may exist, nature of work, and other considerations.

JHAs and JSAs can be conducted differently and mean different things to different organizations. And, while performing a JHA is one tool that can be effective, it is not the only way to identify hazards. For that matter, if they are relied upon too much, and are prepared in a non-participative and pencil-whipping way, effectiveness can be reduced.

Indeed, hazard identification and the assessment of risks from hazards are requirements for organizations wishing to conform to non-regulatory safety management standards like the ANSI/ASSP Z10 and ISO 45001. It is also sometimes required in contracts. The decision to include the assessment of "risk" in an organization's safety program is discussed in more detail in Chapter 5.

Adding the assessment of risk, i.e., estimating the risk that can stem from identified hazards, can help organizations know where to focus efforts and to make risk-based decisions. This is often done in a qualitative or semi-qualitative way, fed into the organization's risk criteria, and then used as part of broader risk management approaches.

OSHA does sometimes require hazard assessments. For example, OSHA requires organizations to determine PPE requirements at:

> 1910.132(d)(1)
> The employer shall assess the workplace to determine if hazards are present, or are likely to be present, which necessitate the use of personal protective equipment (PPE).

Another example is when an employer is subject to the requirements of federal OSHA standards at 29 CFR 1910.119, Process Safety Management of Highly Hazards Chemicals. This standard requires organizations to accomplish a process hazard analysis (PHA) of systems subject to coverage. The standard's purpose is to "prevent or minimize the consequences of catastrophic releases of toxic, reactive, flammable, or explosive chemicals."[17] However, a PHA is conceptually and in practice different from a JHA. The PHA is systems-oriented and applies to hazardous chemical processes somewhat broadly, but still requires analysis of what could go wrong. OSHA accepts several methods of accomplishment for that analysis (like a Hazard and Operability Study [HAZOP], What-If Analysis, or others), which are broader and more technically complicated than a JHA. The completion of JHAs can be complementary, but they look more at the physical tasks of workers.

The bottom line is that hazard identification, such as through the completion of JHAs or other methods, can be used to help identify issues that the organization should control. There are other supplemental ways organizations can identify potential hazards, too, such as through safety walkarounds and safety team inspections.

JHAs certainly aren't new. They are focused on task-level hazards and controls, and as such can be part of an organization's bigger-picture approach to risk management. A full discussion of risk management concepts is beyond the scope of this book. There are several books that discuss risk management. The current construct of risk management, in the context of current standards, is that risk management includes risk identification, risk analysis, and risk evaluation. Excellent coverage of the topic is provided in a technical report published by ASSP.[18] ASSP TR-31010-2020 provides a strategic approach to risk management that covers multiple organizational levels and system life cycles.

One other application of the concept is to apply hazard and risk analysis techniques when new processes, facilities, work cells, and other things are in the design phase, to try to design out hazards before they ever exist. This is sometimes referred to as "Prevention through Design" (PtD), which we will discuss in Chapter 11. Much more information on that is available through the ANSI/ASSP Z590.3 standard. Consider making the implementation of a PtD process part of the requirements you establish for your organization.

Organizational Causes and Considerations

Just as whether you have a safety program or not can impact safety in your organization, the way your safety program is written and the way your organization administers it can either diminish or multiply the effectiveness of your program.

Organizational specifics and business environment, management styles, and methods can all impact safety on the job, too. For example, two different companies could have exactly the same written programs and similar work, yet have significantly different levels of safety and safety outcomes (injury experience). Take ladder safety. A written program on that topic could read almost the same for the two companies.

One reason for differences in safety outcomes could stem from business practices and how different parts of the business communicate and work with each other. How well they plan and make the right ladders available could make a difference.

Another reason for different outcomes could stem from the respective organizations' understanding and beliefs about what causes incidents, culture, and risk tolerance levels. Culture impacts how safety programs are implemented and how work gets done. Company culture could emphasize skipping inspections or doing work in too much of a hurry or improper climbing practices. As a result, work may then be done in either a safer or a less safe way.

One of the larger points here is that the effectiveness of programs can vary; readers should not plan to simply create their written programs and then assume they will run on auto-pilot. Writing a program but then failing to administer it or ignoring it can be almost as bad as not having a program at all. I suppose it could even be worse, since workers could know a program exists but that it is not enforced. For some organizations, that could have a "Wild West" effect.

When you create your written program, you should think about how to best implement it and make it work in the real world.

Organizational safety systems and implications, including managing for "the whole" as opposed to only siloed or individual parts, are discussed in Chapter 8.

Accident/Incident Causation Theory: Beyond "Getting Too Close"

Beyond the injury causes that we can infer from OSHA requirements and what the other industry data and investigation can yield for us – you may want to know that opinions about the *underlying* causes of injuries and about what to emphasize in prevention programs have changed over time. Here is a little background. You may want to consider some of the newer information and concepts, as opposed to stopping at the "unsafe acts and unsafe conditions" level.

Other and newer approaches to injury prevention and thinking about what causes incidents include taking routine proactive action to identify hazards and control them, thinking in a different way about incident causes and drivers, and approaching safety management differently.

People have certainly thought about bigger-picture reasons for why incidents have occurred in the past. Researchers have, for example, analyzed human factor reasons for error (e.g., someone expected a switch moving upward to turn something on instead of off), other reasons why people make errors (assumptions, a variety of potential biases, and other things), engineering-based systems safety, component reliability, and engineering controls.

However, there is more emphasis now on how humans interact with the systems they work with, how people think and act, and organizational systems overall. There is a greater understanding of the frequency of human errors, and evolution of thought about how people make decisions that can impact safety, as well as how organizations and people make trade-offs between efficiency and thoroughness.[19] We can find a multitude of opinions and approaches by researchers and consultants.

I see agreement on many of the factors and hazards that can cause injury, although there is no "universal theory." But we don't have`to go there, and that is not my intent.

It is only possible here to mention some of the discussion, in a broad way. If something piques your interest, consider reviewing some of the many publications on this topic in more detail, and consider taking some focused safety classes and conferences. Keep in mind that sometimes opinions about safety frameworks can be strong. That doesn't mean that there can't be multiple pathways to get to where you want to go. But some models and approaches may be more useful to your program and injury prevention efforts than others.

The primary point is that theories and focus areas for prevention have changed over time. Some approaches are used more in certain industries than others.

To the extent that there are *factors influencing occupational safety* that aren't already spoken to by OSHA requirements, shouldn't we think about them as we develop or revise our safety programs? We have a lot of hints that can suggest approaches to management.

I encourage you to consider a variety of potential injury causes and hazards when you develop your safety programs (written and otherwise). Plan to think beyond macro classification of unsafe actions and unsafe conditions. Plan to think about a broad range of hazards, including nontraditional ones, and about how there are many things that can influence how effective your safety program is.

I also encourage you to consider human performance factors and the actual or de facto management system your organization has in place – things that impact decisions and actions – when interpreting and applying your program's requirements. How your organization engages with its workers can be just as or more important as what you have on paper. However, your organization *must* still have certain things on paper to satisfy the requirements of safety and health regulations.

Conclusion

Organizations should consider a range of things when developing an OSH program. That includes what is known to have the potential to cause injuries and illnesses, but also potential hazards that could be identified by using hazard identification tools or in other ways. Research and talk with other knowledgeable people (as needed) to learn more about hazards and risks that pertain to your operations. Many things can impact occupational safety besides the inherent risk of processes and tasks that workers use or are exposed to. Consider including things in your program that go beyond the obvious. Think about developing your plans in consideration of not only what applicable safety regulations directly require, but also what your organization wants to "make happen" to create an effective framework and processes for safety management.

Notes

1 Extracted from the National Safety Council website (nsc.org), July 2, 2024.
2 The term "near hit" is intended here to mean the same thing as what others may call a "near miss." They are both meant to describe an incident or circumstance where something transpired that could have resulted in injury or property damage if things were just slightly different, but did not. An example could include a tool falling from elevation where someone was nearby but not close enough to get hit by the tool. Different organizations may have nuanced definitions, but fundamentally we are talking about cases where something bad could have happened but didn't, all for what might appear at face value to be a function of chance.
3 The term "unsafe condition" refers to physical or environmental conditions that can result in some degree of risk of harm, such as an injury or illness, if people are exposed to that condition For example, working on an oily floor can be an unsafe condition that could result in a slip or fall. Working near equipment with sharp edges could represent an unsafe condition due to the potential to be cut. (Of course, generically speaking, when we know of hazards that people could be exposed to, we want to fundamentally and preferentially eliminate those hazards or, if impossible or infeasible, apply control measures to minimize injury and illness potential.)
4 Consistent with the definition of "hazard" provided in ANSI/ASSP Z10.0–2019. American Society of Safety Professionals. (2019). *Occupational Health and Safety Management Systems* (ANSI/ASSP Z10.0–2019). American National Standards Institute.
5 This is not your typical list. There are a lot of injury-related discussions that you can find in safety podcasts and various publications. This is, instead, my translation to you, intended to consolidate some bottom-line points. There are some overlapping concepts. You can find much more information about any of these things elsewhere if you investigate further (although you may see nuances in language).
6 With that said – that it can be helpful to have written descriptions of organizational expectations and processes even when OSHA does not require it for a particular situation – there are also "balance" and potential liabilities to be considered here. Opinions can vary, but organizations may want to be careful (again, referring to situations where OSHA does not require a written program or plan) to not establish processes and requirements that that would be difficult or impossible to fulfill.
7 Conklin, T.E. (2019). *The 5 Principles of Human Performance: A Contemporary Update of the Building Blocks of Human Performance for the New View of Safety.* Independently published.
8 Product liability can be a broad topic and is not in the scope of this book.

9 American National Standards Institute, American Society of Safety Professionals, and International Organization for Standardization. (2021). *Occupational Health and Safety Management – Psychological Health and Safety at Work – Guidelines for Managing Psychosocial Risks* (ANSI/ASSP/ISO 45003-2021). American Society of Safety Professionals.

10 Retrieved April 24, 2025, from https://www.cdc.gov/niosh/learning/safetyculturehc/module-2/8.html.

11 As announced at the National Safety Congress, September 2024.

12 One example is: Hagan, P.E., Montgomery, J.F., and O'Reilly, J.T. (Eds.) (2015). *Accident Prevention Manual for Business and Industry, Engineering and Technology* (14th ed.). National Safety Council.

13 Popov, G., Lyon, B.K., and Hollcroft, B.D. (2022). *Risk Assessment: A Practical Guide to Assessing Operational Risks* (2nd ed.). John Wiley & Sons.

14 Lyon, B.K., and Popov, G. (2021). *Assessing and Managing Risk: An ERM Perspective.* American Society of Safety Professionals.

15 https://www.aiha.org/get-involved/volunteer-groups/volunteer-committees-bodies-of-work/industrial-hygiene-hazard-identification-and-exposure-risk-assessment-by-market-segment.

16 Popov, G., Lyon, B.K., and Hollcroft, B. (2016). *Risk Assessment: A Practical Guide to Assessing Operational Risks.* John Wiley & Sons.

17 Process Safety Management of Highly Hazardous Chemicals, 57 Fed. Reg. 6356 (February 24, 1992).

18 If you have a deeper interest, consider seeing the ASSP's report on risk management techniques. American Society of Safety Professionals. (2020). *ASSP TR-31010-2020: Risk Management – Techniques for Safety Practitioners.* American Society of Safety Professionals.

19 Hollnagel, E. (2009). *The ETTO Principle: Why Things That Go Right Sometimes Go Wrong.* Ashgate.

5 Program Elements and Structure

Intent for Chapter

This chapter is intended to provide you with an overview of some of the things that organizations[1] need to have written occupational safety and health (OSH) programs and plans for, if the organization is covered by (regulated under) the Occupational Safety and Health Administration's (OSHA) regulations that apply to general industry or construction.[2] It also mentions some of the things that an employer *may* need, depending upon what types of processes it has, what types of tasks might be performed, and whether or not its processes can result in occupational exposure to particular types of materials or atmospheres.

It further describes a few things to consider that OSHA does not require to be written, but that you may still want to spell out in documents to help make processes or expectations clear to the organization.

It is of course not possible to describe all federal, state, and local requirements, nor all nuances of application within this short book. Instead, the chapter provides information that you can use or consider as you work toward establishing what you required to create and what, beyond that, you would like to create for your organization. Specific needs and requirements will always depend upon your organization's location, type of business, processes, and context.

Safety Program – Background and Levels of Detail

Different organizations may have different requirements and reasons for taking the approaches they choose when deciding what to include in their written programs.

Regulatory requirements will vary based on organizational missions and processes and are usually the major driver in what organizations build into their safety programs. I'll cover more about that in a moment. Beyond that, know that other plans and procedures may be needed to describe the non-mandated things it would like or needs to see, based on internal or other requirements, such as a parent company's or owner's desired work processes. Doing so can be necessary and appropriate. However, it is important to be careful in determining how that other content is written.

DOI: 10.4324/9781032704586-5

Some feel it is important to write very long manuals about how they want processes to work, while others take a minimalist approach. Note that organizations don't get extra credit from anyone for writing long manuals. Lengthy plans and procedures (when unnecessary) create confusion and disinterest. On the flip side, incredibly short programs may not meet OSHA requirements for content, and beyond that often don't have enough detail to be useful to anyone.

In my opinion, it's helpful to have enough information to make clear what the organization requires and expects, and to tie together various parts of a program. However, written programs and plans should not try to describe requirements that go beyond OSHA requirements without being sure that those other things are reasonable and feasible to accomplish sustainably. Documents should describe what is truly expected and what the organization will support and provide resources for. In other words, the organization should plan to do what it says it will do, provide necessary resources, avoid overcommitment, and take steps to ensure that the organization has processes in place to accomplish what is planned. Even though we might like to provide a lot of details, it is important that whatever we develop can be understood easily and is useful in practice.

If an organization sets "pie in the sky" or ambiguous expectations that it does not achieve, or there are program elements that are ignored intentionally through program drift over time, then people may ask, "what else is the organization not doing?" It can also send mixed messages to workers.

Many of us have heard of the proverbial four- or five-inch-thick safety binder that sits on someone's shelf unread, or the electronic version that is lost somewhere in the company's shared drive. Those situations don't do anyone any good. It does not impress inspectors or auditors to point to binders and folders that no one remembered until after an inspection started and which took subsequent hours to locate. Believe it or not, I've seen procedures that sat on shelves for years that were not only not used but that still had blank sections that were intended to list responsible people. Yes, the organization had purchased a third-party fill-in-the-blank program, and then did not tailor it to fit the organization and make it a useful document.

Procedures that are left like that, or on the flip side describe impossible tasks and contradictory requirements, create situations where employees don't know what the real requirements are. That can result in no action, unintentional wrong actions, or unfortunately even egregiously compliant actions. What are the organization's real expectations if they are not written clearly?

As a different but still higher-level issue, it is suggested that organizations write program documents in a way that provides workers with a means to handle situations that are "unexpected" or out of the ordinary.

What will your plans and procedures say that employees and volunteers are expected to do in response to changing circumstances, new or changed hazards, or a loss in the ability for existing controls to adequately control hazards? This is especially important for remote and field work. The issue may be partially addressed through the use of "stop work authority" and "management of change" programs, both of which are described later in this chapter.

Consult with an appropriate attorney if you have legal questions and nuances that need to be considered. Legal staff may have preferred approaches to the management of certain parts of safety programs, such as for auditing processes, and the handling of inspection and corrective action information.

OSHA Requirements – Where Can You Find Them?

The OSH Act itself, the Act that created OSHA to begin with, is located in United States Code Sections 651–678. The Act is the enabling law and has some but not all of what you need to see (including the General Duty Clause). You could look up the Act directly through www.uscode.house.gov. You could also go to the website of the Cornell Legal Information Institute at www.law.cornell.edu. However, that's probably not what most readers need directly in our context for writing procedures, and the Act can be easily accessed through the OSHA website at www.osha.gov.

Federal OSHA's regulations applicable to general industry and construction employers, as well as maritime and agriculture employers, are published in the Code of Federal Regulations (CFR). Those are available online through www.eCFR.gov. However, unless you need to see the full CFR or look up prior versions of regulations, there is an easier way.

The easier way is to go to www.osha.gov. OSHA's standards are well organized there and published alongside other very helpful resources. OSHA's regulations are located in what is referred to as Title 29 of the Code of Federal Regulations. The parts of that most applicable to us for this book are 1900–1910 and 1926. What are thought of as the general industry standards and the construction standards are correspondingly referred to as 29 CFR Part 1910 or 29 CFR 1910 for short, and 29 CFR Part 1926, or 29 CFR 1926 for short.

Employers that operate in state program ("State Plan") states can find their state-specific regulations through links off the main www.osha.gov page, or at www.OSHA.gov/stateplans.

If looking up federal standards from the osha.gov website, click on "Standards" on the top navigation bar, then "Law and Regulations." Once that page opens, under the column titled "Standards," you can choose the industry that you'd like to see the standards for, e.g., general industry or construction. You could also select from other options, including the recordkeeping regulations, which are not located in Parts 1910 or 1926, but rather Part 1904.

Note that the structure of the regulations for State Plans can be different than for the federal site. This is in part because some State Plans had requirements before the OSH Act, plans are built into state law structures, and so on. Also note that State Plans are required to be at least as effective as federal OSHA to be eligible to be an approved plan. Sometimes the State Plans simply adopt the Federal standards, but there are many cases where the state-specific standards are different than the federal, both in structure and in requirements.

While the list of federal regulations can be viewed easily in www.osha.gov, reading through them in larger chunks is not always easy because of the page views. If that is a problem, you can also print sections, order the paper versions from the

Government Printing Office or other sources, view the standard through a commercial safety-related subscription service, or go to www.eCFR.gov for a different view.

Where OSHA Requires Written Programs, Plans, and Procedures

Within the www.osha.gov website, if you click on "Standards," then "Laws and Regulations," then scroll down under "Standards" (yes, you see it again), and then click on the link for "General Industry," or "Construction," a new page will open. The page that opens will provide an index to the category of standards you selected (either 1910 or 1926). You can then review that index, find the name of a particular standard that you are interested in, and click on its link to open the standard. This is where you can read the scope and requirements of standards to determine if your organization will explicitly need a written program, plan, or procedure. You can also determine what other things you might need to make happen but for which OSHA does not require a written program.

You can see by reviewing some of these that to have a "safety program" means more than to have a mere inspection sheet, which is what many people seem to (initially) think.

Regulations can change, but for the federal regulations this can be a starting point for obtaining details on when written documents are mandated.

In general, employers may need a written "something" (plan, program, procedure) based upon any of several things. One could categorize these conceptually to include:

- The types of activities it might be engaged in (e.g., asbestos abatement, hazardous waste site activities)
- Processes and equipment it might have (e.g., mechanical presses, chemicals involved in processing exceeding certain quantities)
- Locations of work (e.g., at locations with fall potential, or confined spaces)
- Potential exposure to the unexpected release of hazardous energy
- Exposures to specific types of environments (e.g., confined spaces with atmospheric hazards) or hazardous materials (e.g., working in processes and areas with potential exposure to chemicals with substance specific standards, such as lead and respirable crystalline silica)
- Other situations

However, it is also the case that even when there is no mandate for a particular plan, OSHA requires certain things. The way it describes its requirements can sometimes get tricky. For example, the general industry forklift standard does not require employers operating powered industrial trucks to implement an overarching written forklift safety "plan." However, it does require that operators be trained and deemed competent to operate the equipment. It also describes requirements for a training program with mandatory content. So it requires a program, but specific to training.

As another example, employers have an obligation to protect workers from recognized hazards when there is no explicit topical standard, under the General Duty Clause 5(a)(1). There are some other things that have to be true in order for OSHA to cite an employer under 5(a)(1),[3] but the point is that an employer has to do *something* to be sure it is protecting its workers from these types of hazards. While OSHA does have some requirements for inspections for various context-specific things, there is no requirement for an employer to have a comprehensive, across-the-board hazard assessment program. Practically speaking, this is part of what drives many organizations to have a formal hazard assessment program, beyond what is required for defined situations (like for construction site inspections described under 29 CFR 1926.20(b)(2)).

When OSHA *does* mention a plan or program, the way that OSHA describes its requirements varies between standards, as can the scope of inclusion.

Tables 5.1 and 5.2 provide a summary of requirements for written programs or plans required by the federal standards.[4] This is intended to provide some ideas for you, subject to your own research and verification.[5] If your operation is in a state subject to an approved State Plan, then this information could be helpful, but you/your organization will also need to review the state's specific requirements.

State requirements can vary widely. Information about State Plans and their requirements and links to State Plan websites can be found at https://www.osha.gov/stateplans.

One state I am aware of has a law that restricts it from adopting anything other than the federal OSHA regulations, and many other states basically adopt federal requirements verbatim. Some states adopt the federal standards but then add their own, or adopt the federal standards but make some changes here and there. Still other states have their own and significantly different (but equally effective!) rules. States that are often thought of as going well beyond the federal, at least for certain standards, include California, Michigan, Oregon, and Washington.

While I am not attempting to specify here all states' requirements that may differ from federal OSHA requirements, due to structural differences and breadth, some differences include:

- Several states require an injury and illness prevention plan (IIPP or I2P2), or similar. Other plan names include accident prevention plan, written safety plan, and workplace accident and injury reduction (AWAIR) program. Sometimes a plan like this is required of all covered employers, and sometimes it depends on other factors, such as employer size or whether the employer is considered "high hazard."
- Examples of states having additional written plan requirements include Tennessee OSHA requiring exposure control plans (under the bloodborne pathogens standard) to include a needlestick prevention plan. Another is North Carolina having a communications tower standard that requires written plans in certain instances. Yet another is California OSHA having requirements for written plans in the areas of workplace violence protection, heat stress prevention, and aerosol transmissible diseases (ATDs).

Table 5.1 General Industry Standards Written Program Requirements

Subject	*Standard Number*	*When*	*Written?*	*Review/ Update*	*Contents*
Emergency Action Plans	1910.38	Evacuation due to emergencies	Yes, unless there are 10 or fewer employees	As necessary	1. Emergency escape procedures 2. Escape route assignments 3. Procedures for critical plan operations shutdown 4. Procedures to account for employees 5. Rescue and medical duties 6. Means of reporting fires and other emergencies 7. Names and job titles of contact person
Hearing Conservation	1910.95	Exposure above 85 dBA for 8 hour TWA	No	As necessary	1. Monitoring of noise levels 2. Annual audiometric testing 3. Hearing protection 4. Training
Process Safety Management	1910.119	1. Presence of chemical above the threshold quantity 2. Presence of flammable chemical or gas in excess of 10,000 lbs 3. Manufacture of explosives or pyrotechnics	Yes	As necessary	1. Procedures for: 2. Employee participation 3. Compilation of process safety information 4. Process hazard analysis 5. Operating procedures 6. On-going integrity of process equipment 7. Management of change 8. Incident investigation 9. Emergency planning and response

(Continued)

Table 5.? (Continued)

Subject	*Standard Number*	*When*	*Written?*	*Review/ Update*	*Contents*
Emergency Response Plans	1910.120	Possibility of uncontrolled release of a hazardous chemical	Yes	As necessary	1. Pre-emergency planning 2. Coordination with outside parties 3. Personnel roles 4. Emergency recognition and prevention 5. Safe distances and places of refuge 6. Site security and control 7. Evacuation routes and procedures 8. Decontamination 9. Emergency medical treatment procedures 10. Emergency alerting and response procedures 11. Critiques of response and follow-up 12. PPE and emergency equipment
Respiratory Protection	1910.134	Respirators are required to be worn; overexposure, possible overexposure, unknown atmosphere	Yes, unless is voluntary use of filtering facepiece	Regularly	Procedures for: 1. Selection 2. Medical evaluations 3. Fit testing 4. Proper use in routine situations and emergencies 5. Maintenance, and schedule 6. Ensuring air quality, quantity, and flow for atmosphere-supplying respirators 7. Training: hazards and proper use 8. Regular evaluations of effectiveness
Permit-Required Confined Spaces	1910.146	Employees will enter permit spaces	Yes	Within 1 year of each entry or annually	Procedures for: 1. Preventing unauthorized entry 2. Identify and evaluating hazards 3. Safe permit entry operations 4. Provision of equipment 5. Evaluating permit space conditions 6. Provision of attendant(s) 7. Attending multiple spaces is applicable 8. Identifying authorized entrants, attendants, entry supervisors, persons who monitor, etc. 9. Summoning rescue and emergency services

Subject	*Standard Number*	*When*	*Written?*	*Review/ Update*	*Contents*
Permit-Required Confined Spaces					and rescuing entrants 10. Preparation, issuance, use, and cancellation of entry permits 11. Coordinating entry operations where employees or more than one employee is involved 12. Concluding the entry 13. Reviewing entry operations
Lockout/ tagout	1910.147	Unexpected energization or start up of machine or equipment or release of stored energy could cause injury	Yes	At least annually	1. Energy control procedures a. Scope, purpose, authorization, rules, techniques to be used to control hazardous energy b. Means to enforce compliance c. Statement of intended use of the procedure d. Specific procedural steps for shutting down, isolating, blocking, and securing machines or equipment to control hazardous energy e. Specific procedural steps for placement, removal, and transfer of lockout or tagout devices and the responsibility for them f. Specific requirements for testing a machine or equipment to determine and verify effectiveness of lockout/ tagout devices, etc. 2. Employee training 3. Periodic inspections
Fire Brigades	1910.156	When the employer chooses to establish a fire brigade	Yes – Organizational statement		1. Basic organizational structure 2. Type, amount and frequency of training provided to members 3. Expected number of members 4. Function of the brigade

(Continued)

Table 5.1 (Continued)

Subject	*Standard Number*	*When*	*Written?*	*Review/ Update*	*Contents*
Grain Handling Facilities	1910.272	In grain elevators, feed mills, flour mills, rice mills, dust palletizing plants, dry corn mills, soybean flaking operations, and dry grinding operations of soycake	Yes – a. Emergency action plan b. Housekeeping program		1. Emergency action plan – see 1910.38 2. Housekeeping program a. Frequency and b. Methods to reduce accumulations of dust on ledges, floors, equipment, and other surfaces.
Asbestos	1910.1001	a. When the TWA and/ or excursion limit is exceeded *(PELs=8hr TWA of 0.1 fiber/ cc and 30 min excursion limit of 1 fiber/cc)* b. When respirator use is required	Yes – a. Compliance program b. Respirator program	As necessary to reflect significant changes	a. Compliance program How the employer will reduce employee exposure to below the TWA and excursion limit using engineering and work practice controls b. See 1910.134
Vinyl Chloride	1910.1017	a. When exposures exceed the PEL *(PELs=8hr TWA of 1 ppm and 15 min STEL of 5 ppm)* b. When respirator use is required	Yes – a. Compliance program b. Respirator program	Annually	a. Compliance program How the employer will reduce employee exposure to below the permissible exposure limits or to the greatest extent feasible using engineering and work practice controls b. See 1910.134
Inorganic Arsenic	1910.1018	a. When exposures exceed the PEL *(PEL=8hr TWA of 10 μg/m^3)* b. When respirator use is required	Yes – a. Compliance program b. Respirator program	Annually	a. Compliance program 1. Description of each operation in which inorganic arsenic is emitted 2. Engineering plans and studies used to determine methods for controlling exposure 3. A report of the technology considered in meeting the permissible exposure limit 4. Air monitoring data 5. A schedule for implementation of engineering and work practice controls b. See 1910.134

(Continued)

Subject	*Standard Number*	*When*	*Written?*	*Review/ Update*	*Contents*
Lead	1910.1025	a. When exposures exceed the PELexposure limit *(PEL=8hr TWA of 50 μg/m³)* b. When respirator use is required	Yes— a. Compliance program b. Respirator program	At least annually	a. Compliance program 1. Description of each operation in which lead is emitted 2. Description of specific means to be used to achieve compliance 3. A report of the technology considered in meeting the permissible exposure limit 4. Air monitoring data 5. A schedule for implementation of the program including documentation 6. A work practice program 7. An administrative control schedule, if applicable b. See 1910.134
Cadmium	1910.1027	a. When exposures exceed the PEL *(PEL=8hr TWA of 5 μg/m³)* b. When respirator use is required	Yes – a. Compliance program b. Respirator program	At least annually	a. Compliance program 1. Description of each operation in which cadmium is emitted 2. Description of specific means to be used to achieve compliance 3. A report of the technology considered in meeting the permissible exposure limit 4. Air monitoring data 5. A work practice program 6. A written plan for emergency situations b. See 1910.134
Benzene	1910.1028	a. When exposures exceed the PEL *(PEL=8hr TWA of 1 ppm)* b. When respirator use is required	Yes – a. Compliance program b. Respirator program	As appropriate based on most recent air monitoring	a. Compliance program 1. How the employer will reduce employee exposure to below the permissible exposure limit using engineering and work practice controls 2. A schedule for development and implementation of engineering and work practice controls b. See 1910.134

(Continued)

Table 5.1 (Continued)

Subject	*Standard Number*	*When*	*Written?*	*Review/ Update*	*Contents*
Coke Oven Emissions	1910.1029	a. When exposures exceed the PEL *(PEL=8hr TWA of 150 μg/m³)* b. When respirator use is required	Yes – a. Compliance program b. Respirator program	At least annually	a. Compliance program 1. A description of each coke oven operation by battery 2. Engineering plans and other studies used to determine the controls 3. A report of the technology considered in meeting the permissible exposure limit 4. Air monitoring data 5. A schedule for implementation of the engineering and work practice controls b. See 1910.134
Bloodborne Pathogens	1910.1030	Exposure to blood and/ or body fluids; first-aid responder	Yes	At least annually	1. Exposure determination 2. Methods of compliance procedures 3. HBV vaccinations procedures Post-exposure evaluation and follow-up procedures 5. Labeling procedures 6. Training procedures 7. Recordkeeping 8. Evaluation of circumstances of exposure incident 9. Type and brand of device in use when exposure incident occurred
Cotton Dust	1910.1043	a. When exposures exceed the PEL *(PEL=8hr TWA of 200 of lint- free respirable dust μg/m³)* b. When respirator use is required	Yes – a. Compliance program b. Respirator program	When necessary	a. Compliance program 1. A description of each coke oven operation by battery 2. Engineering plans and other studies used to determine the controls 3. A report of the technology considered in meeting the permissible exposure limit 4. Air monitoring data 5. A schedule for implementation of the engineering and work practice controls b. See 1910.134

(Continued)

Subject	*Standard Number*	*When*	*Written?*	*Review/ Update*	*Contents*
1,2-Dibromo-3- Chloropropane (DBCP)	1910.1044	a. When exposures exceed the PEL *(PEL=8hr TWA of 1 ppb)* b. When respirator use is required	Yes – a. Compliance program b. Respirator program	At least annually	a. Compliance program A schedule for implementation of the engineering and work practice controls b. See 1910.134
Acrylonitrile	1910.1045	a. When exposures exceed the PEL *(PEL=8hr TWA of 2 ppm and a 15 min ceiling of 10 ppm)* b. When respirator use is required	Yes – a. Compliance program b. Respirator program	At least annually	a. Compliance program 1. A description of each operation or process resulting in exposure to acrylonitrile above the PELs 2. An outline of the nature of the engineering controls and work practice controls to be applies 3. A report of the technology considered in meeting the PELs 4. A schedule of implementation of engineering and work practice controls b. See 1910.134
Ethylene Oxide	1910.1047	a. When exposures exceed the PEL *(PEL=8hr TWA of 1 ppm and a 15 min excursion limit of 5 ppm)* b. When respirator use is required	Yes – a. Compliance program b. Respirator program	At least annually	a. Compliance program 1. How the employer will reduce employee exposure to below the PELs using engineering controls, work practice controls, and respiratory protection 2. A schedule for periodic lead detection surveys 3. Plan for emergency situations b. See 1910.134
Formaldehyde	1910.1048	a. When exposure to formaldehyde occurs b. When respirator use is required	Yes – a. Hazard communication program b. Respirator program	Regularly	a. See 1910.1200 b. See 1910.134

(Continued)

Table 5.1 (Continued)

Subject	*Standard Number*	*When*	*Written?*	*Review/ Update*	*Contents*
Methyleledaniline (MDA)	1910.1050	a. When exposures exceed the PEL *(PEL=8hr TWA of 10 ppb and a 15 min STEL of 100 ppb)* b. When respirator use is required c. When there is possibility of an emergency involving MDA	Yes – a. Compliance program b. Respiratory program c. Emergency plan	At least annually	a. Compliance program 1. How the employer will reduce employee exposure to below the PELs using engineering controls, work practice controls, and respiratory protection 2. A schedule for periodic maintenance b. See 1910.134 c. Emergency plan 1. Personal protective equipment and clothing provided to those involved in correcting emergency conditions 2. Provisions for alerting and evacuating affected employees 3. Elements in 1910.38 and 1910.39
1,3 Butadiene	1910.1051	a. When exposures exceed the PEL *(PEL=8hr TWA of 1 ppb and a 15 min STEL of 5 ppb)* b. When respirator use is required	Yes – a. Compliance program b. Respiratory program	At least annually	a. Compliance program 1. How the employer will reduce employee exposure to below the PELs using engineering controls, work practice controls, and respiratory protection 2. A schedule for the development and implementation of the engineering controls and work practice controls including periodic leak detection surveys b. See 1910.134
Hazard Communication	1910.1200	Exposure to hazardous chemicals	Yes	As necessary	1. List of hazardous chemicals 2. Labeling/forms of warning of hazardous chemicals 3. Safety Data Sheets/accessibility 4. Information and training 5. Methods used to inform employees of hazards of non-routine tasks (including unlabeled pipes) 6. Multi-employer activity procedures (access +)

(Continued)

Subject	*Standard Number*	*When*	*Written?*	*Review/ Update*	*Contents*
Occupational Exposure to Hazardous Chemicals in Laboratories	1910.1450	When employees use hazardous chemicals on a "laboratory scale"	Yes – Chemical hygiene plan	At least annually	1. Standard operating procedures 2. Criteria used to determine and implement control measures to limit exposure to hazardous chemicals 3. Procedures to ensure proper and adequate performance of fume hoods and other protective equipment 4. Circumstances under which a particular laboratory operation, procedure, or activity shall require prior approval before implementation 5. Provisions for medical consultation and medical examinations 6. Designation of personnel responsible for implementation of the plan 7. Assignment of a Chemical Hygiene Officer 8. Provisions for additional employee protection for working with particularly hazardous substances

Table 5.2 Construction Standards with Written Program Requirements

Subject	*Standard Number*	*When*	*Written?*	*Review/ Update*	*Contents*
Emergency Action Plans	1926.35	Evacuation due to emergencies	Yes, unless there are 10 or fewer employees	As necessary	1. Emergency escape procedures 2. Escape route assignments 3. Procedures for critical plan operations shutdown 4. Procedures to account for employees 5. Rescue and medical duties 6. Means of reporting fires and other emergencies 7. Names and job titles of contact person
Hazard Communication	1926.59	Exposure to hazardous chemicals	Yes	As necessary	1. List of hazardous chemicals 2. Labeling/forms of warning of hazardous chemicals 3. Safety Data Sheets/accessibility 4. Information and training 5. Methods used to inform employees of hazards of non-routine tasks (including unlabeled pipes) 6. Multi-employer activity procedures (access +)
Methylenedianiline (MDA)	1926.60	a. When exposures exceed the PEL *(PEL=8hr TWA of 10 ppb and a 15 min STEL of 100 ppb)* b. When respirator use is required c. When there is possibility of an emergency involving MDA	Yes – a. Compliance program b. Respiratory program c. Emergency plan	At least annually	a. Compliance program 1. How the employer will reduce employee exposure to below the PELs using engineering controls, work practice controls, and respiratory protection 2. A schedule for periodic maintenance b. See 1926.103 c. Emergency plan 1. Personal protective equipment and clothing provided to those involved in correcting emergency conditions

(Continued)

Subject	*Standard Number*	*When*	*Written?*	*Review/ Update*	*Contents*
					2. Provisions for alerting and evacuating affected employees 3. Elements in 1910.38 and 1910.39
Lead	1926.62	a. When exposures exceed the PEL exposure limit *(PEL=8hr TWA of 50 $\mu g/m^3$)* b. When respirator use is required	Yes a. Compliance program b. Respirator program	At least annually	a. Compliance program 1. Description of each operation in which lead is emitted 2. Description of specific means to be used to achieve compliance 3. A report of the technology considered in meeting the permissible exposure limit 4. Air monitoring data 5. A schedule for implementation of the program including documentation 6. A work practice program 7. An administrative control schedule, if applicable 8. Description of arrangements made among contractors to inform affected employees of potential exposure and compliance responsibility b. See 1926.103
Process Safety Management	1926.64	1. Presence of chemical above the threshold quantity 2. Presence of flammable chemical or gas in excess of 10,000 lbs 3. Manufacture of explosives or pyrotechnics	Yes	As necessary	Procedures for: 1. Employee participation 2. Compilation of process safety information 3. Process hazard analysis 4. Operating procedures 5. Ongoing integrity of process equipment 6. Management of change 7. Incident investigation 8. Emergency planning and response

(Continued)

Table 5.2 (Continued)

Subject	*Standard Number*	*When*	*Written?*	*Review/ Update*	*Contents*
Emergency Response Plans	1926.65	Possibility of uncontrolled release of a hazardous chemical	Yes	As necessary	1. Pre-emergency planning 2. Coordination with outside parties 3. Personnel roles 4. Emergency recognition and prevention 5. Safe distances and places of refuge 6. Site security and control 7. Evacuation routes and procedures 8. Decontamination 9. Emergency medical treatment procedures 10. Emergency alerting and response procedures 11. Critiques of response and follow-up 12. PPE and emergency equipment
Respiratory Protection	1926.103	Respirators are required to be worn; overexposure, possible overexposure, unknown atmosphere	Yes, unless there is voluntary use of filtering facepiece	Regularly	Procedures for: 1. Selection 2. Medical evaluations 3. Fit testing 4. Proper use in routine situations and emergencies 5. Maintenance and schedule 6. Ensuring air quality, quantity, and flow for atmosphere-supplying respirators 7. Training: hazards and proper use 8. Regular evaluations of effectiveness
Fall Protection	1926.502	When engaged in leading edge work, precast concrete erection work, or residential	Yes	Must be "up to date" and specific to the site	1. Reasons why use of conventional fall protection equipment is infeasible or would create a greater hazard 2. Discussion of other measures to be taken to reduce or eliminate fall hazards

(Continued)

	Number			Update	
		construction and can demonstrate that it is infeasible or creates a greater hazard to use conventional fall protection equipment			3. Identity of each location where convention fall protection methods cannot be used and designation of those areas as controlled access zone 4. When no other measure has been implemented, the implementation of a safety monitoring system 5. Names or other method of identification for each employee who is designated to work in a controlled access zone
Vinyl Chloride	1926.1117	a. When exposures exceed the PEL *(PELs=8 hr TWA of 1 ppm and 15 min STEL of 5 ppm)* b. When respirator use is required	Yes – a. Compliance program b. Respirator program	Annually	a. Compliance program How the employer will reduce employee exposure to below the permissible exposure limits or to the greatest extent feasible using engineering and work practice controls b. See 1926.103
Inorganic Arsenic	1926.1118	a. When exposures exceed the PEL *(PEL=8hr TWA of 10 $\mu g/m^3$)* b. When respirator use is required	Yes – a. Compliance program b. Respirator program	Annually	a. Compliance program 1. Description of each operation in which inorganic arsenic is emitted 2. Engineering plans and studies used to determine methods for controlling exposure 3. A report of the technology considered in meeting the permissible exposure limit 4. Air monitoring data 5. A schedule for implementation of engineering and work practice controls b. See 1926.103
Cadmium	1926.1127	a. When exposures exceed the PEL *(PEL=8hr TWA of 5 $\mu g/m^3$)* b. When respirator use is required	Yes – a. Compliance program b. Respirator program	As often as necessary	

(Continued)

Table 5.2 (Continued)

Subject	*Standard Number*	*When*	*Written?*	*Review/ Update*	*Contents*
Vinyl Chloride	1910.1017	a. When exposures exceed the PEL *(PELs=8hr TWA of 1 ppm and 15 min STEL of 5 ppm)* b. When respirator use is required	Yes – a. Compliance program b. Respirator program	Annually	a. Compliance program How the employer will reduce employee exposure to below the permissible exposure limits or to the greatest extent feasible using engineering and work practice controls b. See 1926.103
Inorganic Arsenic	1910.1018	a. When exposures exceed the PEL *(PEL=8hr TWA of 10 μg/m³)* b. When respirator use is required	Yes – a. Compliance program b. Respirator program	Annually	a. Compliance program 1. Description of each operation in which inorganic arsenic is emitted 2. Engineering plans and studies used to determine methods for controlling exposure 3. A report of the technology considered in meeting the permissible exposure limit 4. Air monitoring data 5. A schedule for implementation of engineering and work practice controls b. See 1926.103
Lead	1910.1025	a. When exposures exceed the PEL exposure limit *(PEL=8hr TWA of 50 μg/m³)* b. When respirator use is required	Yes – a. Compliance program b. Respirator program	At least annually	a. Compliance program 9. Description of each operation in which lead is emitted 10. Description of specific means to be used to achieve compliance 11. A report of the technology considered in meeting the permissible exposure limit 12. Air monitoring data 13. A schedule for implementation of the program including documentation

(Continued)

Subject	*Standard Number*	*When*	*Written?*	*Review/ Update*	*Contents*
					14. A work practice program 15. An administrative control schedule, if applicable b. See 1926.103
Cadmium	1926.1127	a. When exposures exceed the PEL *(PEL=8hr TWA of 5 μg/m^3)* b. When respirator use is required	Yes – a. Compliance program b. Respirator program	At least annually	a. Compliance program Engineering, work practice controls, and appropriate respiratory protection to be used to reduce exposures to below the PEL b. See 1926.103
Benzene	1910.1128	a. When exposures exceed the PEL *(PEL=8hr TWA of 1 ppm)* b. When respirator use is required	Yes – a. Compliance program b. Respirator program	As appropriate based on most recent air monitoring	a. Compliance program 1. How the employer will reduce employee exposure to below the permissible exposure limit using engineering and work practice controls 2. A schedule for development and implementation of engineering and work practice controls b. See 1926.103
Coke Oven Emissions	1926.1129	a. When exposures exceed the PEL *(PEL=8hr TWA of 150 μg/m^3)* b. When respirator use is required	Yes – a. Compliance program b. Respirator program	At least annually	a. Compliance program 1. A description of each coke oven operation by battery 2. Engineering plans and other studies used to determine the controls 3. A report of the technology considered in meeting the permissible exposure limit 4. Air monitoring data 5. A schedule for implementation of the engineering and work practice controls b. See 1926.103

(Continued)

Table 5.2 (Continued)

Subject	*Standard Number*	*When*	*Written?*	*Review/ Update*	*Contents*
1,2-Dibromo-3-Chloropropane (DBCP)	1926.1144	a. When exposures exceed the PEL *(PEL=8hr TWA of 1ppb)* b. When respirator use is required	Yes – a. Compliance program b. Respirator program	At least annually	a. Compliance program A schedule for implementation of the engineering and work practice controls b. See 1910.134
Acrylonitrile	1926.1145	a. When exposures exceed the PEL *(PEL=8hr TWA of 2 ppm and a 15 min ceiling of 10 ppm)* b. When respirator use is required	Yes – a. Compliance program b. Respirator program	At least annually	a. Compliance program 1. A description of each operation or process resulting in exposure to acrylonitrile above the PELs 2. An outline of the nature of the engineering controls and work practice controls to be applied 3. A report of the technology considered in meeting the PELs 4. A schedule of implementation of engineering and work practice controls b. See 1910.134
Ethylene Oxide	1926.1147	a. When exposures exceed the PEL *(PEL=8hr TWA of 1 ppm and a 15 min excursion limit of 5 ppm)* b. When respirator use is required	Yes – a. Compliance program b. Respirator program	At least annually	a. Compliance program 1. How the employer will reduce employee exposure to below the PELs using engineering controls, work practice controls, and respiratory protection 2. A schedule for periodic lead detection surveys 3. Plan for emergency situations b. See 1910.134
Formaldehyde	1926.1148	a. When exposure to formaldehyde occurs b. When respirator use is required	Yes – a. Hazard communication program b. Respirator program	Regularly	a. See 1910.1200 b. See 1910.134

Once again, the table and limited examples of state differences are meant to provide you with some framework. Use them as tools but don't rely solely upon them. Terms can be nuanced, requirements can be nested within the OSHA text, and regulations can change.

The information that you have compiled about your organization's current and planned operations can be compared to lists like the one above, but your employer's operations need to be compared to OSHA regulations directly. If you operate in one state, then you'll need to review the federal regulations if federal OSHA programs apply, or the state's regulations if a State Plan applies. If your organization has operations in more than one state, then it needs to be sure it is compliant with the requirements of each state that it operates in (not just the location a headquarters is located in). Some environmental, health, and safety (EHS) online services have portals through which one can search for differences in certain types of regulations between states.

Regulations in general come with some description of what they apply to, either in a section discussing scopes and applications, or through other texts. You might think that something on the list above pertains to your employer, but you might determine otherwise by reading the standard's scope carefully. Some things listed as requiring a program, plan, or procedure may only require something in writing in particular circumstances. The point is that you will need to do that checking.

By the way, your operational information should also be compared to potential non-OSHA obligations that exist or may be evolving from other regulators. For example, the Environmental Protection Agency (EPA) and state environmental equivalents could have requirements that are related to or even coexist with OSHA requirements. Regarding EPA, that could include new or proposed standards under the Toxic Substances Control Act (TSCA).

Also, remember that an organization can have other requirements under OSHA that do not fall neatly under the standards in 29 CFR 1910 or 29 CFR 1926 or state equivalents. For example, the federal injury and illness recordkeeping requirements are described at 29 CFR 1904, as opposed to 1910 or 1926. Go to https://www.osha.gov/laws-regs/regulations/standardnumber to see a list of federal regulations/standards within 29 CFR. You can use that URL to obtain links to the recordkeeping rules, the general industry regulations, the construction regulations, and many others. You can expect that if you see a regulation on the federal site about a particular topic, but you work in a State Plan state, the state will have a similar regulation that has been determined previously by federal OSHA to be at least as effective as the federal requirements. That's just due to the rules for State Plan approvals by federal OSHA.

The Writing Process

Program, Plan, and Procedure Templates

You may want to know right away that there are several different ways to obtain templates for programs that your organization needs, if you don't want to develop them from scratch. There are several for-fee organizations that provide templates for required programs, plans, and procedures, as well as organizations that provide some for free.

Accessing templates from a fee-based company could help your processes, and the templates can be a great resource. Providers of those templates sometimes also offer additional resources and help as part of a subscription service.

Sometimes associations provide sample programs that can be obtained for free if you are a member of the association. They can be good and fit well with the needs of particular industries and types of work.

There are also other organizations that provide templates and tools for free. As an example, for the construction industry, there is an online tool to develop a respirable crystalline silica exposure control plan, a template for a written fall protection and rescue plan, and other resources that can be obtained through the Center for Construction Research and Training (CPWR) at the https://www.cpwr.com website.

Federal OSHA offers compliance assistance programs and various templates for written programs. OSHA provides many sample programs at https://www.osha.gov/complianceassistance/sampleprograms.

State Plans also have some excellent resources available. For example, California OSHA has many publications, including templates for IIPPs and many others. See https://www.dir.ca.gov/dosh/puborder.asp.

For any sample template that you can find, it may be very generic, written in a way that is too simple or too complex for the location you are assisting, or not fit your culture. It will always be the case that your organization will need to tailor templates to fit the organization and specific work scenarios. Sometimes that can be done very easily, and at other times take more work. Tailoring may be needed, e.g., for job titles, process flow descriptions, process timelines and deadlines, document locations and references, the inclusion of your other location specific work information, and how your organization really wants things to work.

You may also have business acquaintances that have existing programs similar to what you need. Sometimes those acquaintances may be willing to share, and at other times be very secretive about or protective of their documents. In the event that you want to obtain a document this way, make sure that neither you nor your acquaintance are creating any copyright, proprietary, or confidentiality issues. And, of course, just like obtaining templates in other ways, it is up to your organization to be sure that the documents obtained contain all the OSHA-required information and that the processes are tailored to fit.

Writing for Yourself, and Tailoring: Get Input by End-Users

In my opinion, some portions of a safety program might be best written by a safety manager or topic-specific subject matter expert. This can be the case especially for portions developed before a facility has started construction or when the program framework is first being considered.

However, some portions can be well suited to development in consultation with other workers/end-users and other interested parties, such as departments that will need to know or be impacted by the program. This can of course depend on the timing of your work.

Programs need to meet requirements and, at the same time, be realistic. Consultation with others can help you be aware of a variety of issues that you might not have thought about, or to take into account factors that others with more specific experience might have. Have you ever heard something like, "I don't know why we have to do it this way … it makes no sense – but then again, nobody asked me." Plus, consulting with others can increase buy-in and compliance with the ultimate requirements.

General Structure and Format Considerations

Federal OSHA does not define how you name or structure particular programs. It does specify elements of content. As mentioned before, OSHA uses different terms to sometimes mean essentially the same thing.

Federal OSHA does use the terms program, plan, and procedure. In common use, "program" most often describes an organization's comprehensive approach to something, "plan" usually refers to how an organization will approach or work through a particular issue or type of work, and a "procedure" usually refers to chronological steps to do something. I sometimes refer to them collectively as "requirements documents" for convenience, because each of them references portions of an organization's OSH requirements.

State Plans, however, sometimes require employers to have written documents referred to by different acronyms, such as IIPP (injury and illness prevention plan). As for federal OSHA written program requirements, content is specified. If a state requires an IIPP or equivalent, it will also specify mandatory content.

Some organizations structure their OSH programs in multiple tiers. OSHA does not specify that this is needed. Some organizations need something that is complex, while others have no need for a complex approach and in fact would do best to keep things very simple.

When organizations use tiers, that can be a function of company size, complexity, contract requirements, a need for multiple types of delineation, or other things.

For those who might need it, a multitiered structure can include having:

- A program, or occupational health and safety management system

  A program could sometimes be meant to describe an overall approach to a narrow type of safety issue, e.g., perhaps a forklift safety "program."

However, many employers refer to their safety "program" as the sum total of their approach to handling a variety of hazards. In this case their "program" is an assembly of approaches to individual safety issues. For example, it could consist of a forklift safety program together with a welding safety program together with a hazard communication program, and so on.

Sometimes an employer may purposefully create a safety management system, or an occupational health and safety management system (OHSMS). An OHSMS basically addresses individual types of hazards and controls but also describes how the organization will manage its overall processes and the relationships between processes. One of the intents for using an OHSMS is to have an overall process that is better in safety results than the sum of the parts would be. There is more information about OHSMS, and the rationale for using them, in Chapter 8.

- Policies:

  A safety "policy" is often a one-page and high-level document placed at the front of a collection of safety requirements documents (programs, plans, procedures, etc.). When used in that manner, it describes an organization's overall commitment to safety, scope of coverage of requirements documents, and responsibilities. Organizations that decide to apply a particular OHSMS consensus standard will have certain content that is required by the OHSMS (e.g., ISO 45001). It is sometimes signed by the most senior executive. These can sometimes provide an important frame of reference when struggling internally with potentially conflicting internal priorities.

However, sometimes organizations, often smaller ones, will create what they refer to as a safety "policy" of several pages, essentially establishing a set of safety rules. These sets of safety rules often fall short of OSHA requirements, if the company performs types of work for which OSHA requires a written program or plan. Again, it depends. OSHA does not so much care what an organization calls its requirement documents (although it could make it more clear to OSHA what the documents are intended to address). It's more a matter of whether what you describe for requirements contain content required by applicable OSHA standards.

- Procedures, safe work procedures, or standard operating procedures:

  These often describe specific required processes and actions, in a stepwise or progressive fashion. They may include responsibilities and describe the organization's requirements for certain activities. They may contain flow charts to help readers understand intended relationships visually, as a supplement to text. They are sometimes written at a "higher level" about how to ensure things get done safely, or how the organization will fulfill some OSH requirements for compliance. However, sometimes they describe expectations for stepwise workflow. It can depend upon organizational context, size, and culture.

- Plans:
  These often describe how an organization will handle a particular contingency – as in, for example, an "Emergency Action Plan." But not always. Sometimes a "plan" is what OSHA requires an employer to create to describe how it will address a particular type of issue, such as for the control of worker exposures to respirable crystalline silica (RCS).

- Specific requirements:
  Some organizations create more detailed requirements that they want addressed in a certain way. Specific requirements can often contain more technical-oriented or focused requirements than some procedures. Again, it's up to the organization to determine what documents it wants to create, how many levels of procedures it wants, which documents it wants to make mandatory, and so on.

- Guidance documents:
  These are oftenintended to describe possible approaches and organizational preferences, without specifically mandating something. These can be helpful when approaching safety programs located in different parts of the world and country, where regulations may be different or where latitude in approach is necessary.

- Work instructions:
  These generally provide specific instructions for how to accomplish particular work tasks. They would often be the lowest-level document in a requirements document hierarchy, and be quite nitty gritty.

With all this said, many employers will call their requirements documents different things. That can be based on corporate or local preference or just some site history. If inspected by OSHA, the inspector (Compliance Safety and Health Officer [CSHO]) should look for whether a program, plan, or procedures inspected contains required elements, as opposed to what the organization chooses to call a particular document.

There is no OSHA-inherent requirement to have a program that is seven layers deep; what matters is what is effective for your organization.

Regardless of what your organization calls its requirements documents or framework, what is most important is that management and employees understand the structure and requirements as it pertains to them.

Management should understand, at least, the organization's approach to safety and have some understanding of the applicable content. It should understand the organization's framework for approaching OSH generally, its overarching policies, and requirements related to what they are responsible for managing. Other employees should understand especially the policies of the organization, the value that the organization places on safe work and a safe workplace, how to obtain OSH information and documents requirements, the requirements of the processes that are relevant to them, and expectations.

A Few Other Pointers on Requirements Documents

Consider the following:

- Job titles versus names. Unless you are required to do so by a state or local regulation, it may be preferable to use job titles when specifying responsibilities, as opposed to names.[6]

  Depending on the document and the circumstance, the use of names could create issues for effective management. For example, if there is only one person with a particular job title, you might consider referencing the title of the person normally responsible for something but then also leave room for a "designee" in the event someone is not available.
- Requirements documents should be accessible to those who need them. I've seen some organizations write programs but then make them nearly impossible for workers who are supposed to be following them to access. That is not the intent for laying things out in writing. Yes, there can be several ways an organization can provide workers with access to procedures, but if your organization is inspected, don't be surprised to be asked how workers (that includes senior executives as well as people who work on a production line) access the directions that they are supposed to follow.
- Requirements documents should be written in a way that keeps the end-user in mind. The more easily understood, the more likely to be implemented in the manner intended. Strive for balance.

What OSHA Commonly Requires in Written Programs/Plans

As you might guess, you will need to go to each applicable regulation and read what OSHA explicitly requires to be included. There is no single answer. However, to give you a feel for the matter, one could say that OSHA sometimes requires written documents that might be characterized as requiring:

- A description of the scope of a program/plan and what the program/plan is meant to cover, i.e., how coverage will be implemented
- Indication of which positions are responsible for completion of requirements
- Information about potential hazards and requirements for the assessment of hazards (whether physical or industrial hygiene related)
- A description of how employees will be trained to recognize hazards
- What the employer will do to protect employees from hazards
- Steps or procedures needed to control hazards (e.g., exposure control plans, requirements for inspections)

- Communication and training requirements; how the employer is to make the employees aware of the hazards
- Documentation and recordkeeping obligations

The bullet points above are just examples to give you a general feel for the kinds of things that OSHA can expect. You'll sometimes see those types of things and other things described differently, in the assortment of OSHA regulations that require written programs.

Other Supporting Processes and Frameworks

OSHA does not require written procedures for everything your organization needs to do or that your workers need to do. *However*, sometimes it does require that certain things be done, while not mandating that you describe your process in writing.

You have probably seen cases where certain tasks could pose a safety risk and just beg for specific procedures that address things like work methods/instructions, training requirements, and preventive maintenance processes. That could be to help address general hazards, OSHA compliance, or other management system needs.

So, what is an organization to do? Sometimes the requirements are such that "unwritten" processes can be adequate (unless mandated by OSHA). But memories and perceptions can certainly fail sometimes. The adequacy of unwritten processes can change over time. How good is your memory? How about the memory of other managers? With rapidly changing circumstances and a multitude of things to address and often competing priorities, your organization should carefully consider what it wants to put in writing vs what it wants to accomplish by institutional memory.

There are several supporting processes that your organization may need, and which are not or only in some cases required by OSHA. Here are a few processes to consider spelling out.

Hazard Analysis and Inspection Processes

As we have discussed elsewhere, employers have (OSHA) regulatory obligations and other potential liabilities that make it important to have some types of hazard identification and inspection processes. There are specific federal OSHA regulations that require different things, like frequent and regular inspections of "job sites, material and equipment" by competent people on construction projects, inspections for scaffolds, excavations, cranes, walking/working surfaces, powered industrial trucks, and others.

Some State Plans formally require, as part of their IIPP (or similar program) requirements, that hazard identification and evaluation processes be built into their programs.

Therefore, it is sometimes mandated by regulation and otherwise makes sense to provide some framework for your organization regarding how its hazard identification and control processes will work.

For example, pre-use and other equipment and tool-related inspections might be handled one way and accomplished by the workers utilizing them and through other competent third-party organizations. However, I've seen some companies have their cranes, for example, inspected with unacceptable gaps between inspections. And there are many other things that must be happening, and documentation that must be kept. You might want to provide a framework to help ensure that mandatory inspections are happening.

Apart from special purpose equipment or process inspections, employers generally manage broader and more general walk-throughs and inspections differently. Will there be checklists? Will results be shared and tracked? Will deficiencies be forwarded to some department or entered into a work-order system for correction? How should that work?

If you don't define your process to an appropriate level, what gets done will likely be more variable in coverage and effectiveness than if you do.

Yet another type of inspection and hazard analysis process can pertain to new facilities being constructed and new processes/equipment being installed.

Defining desired processes, required reviews, persons involved, and hazard assessment tool options can be helpful. You probably don't want to put your organization into a corner, but outlining expectations and options can be much better than expecting people to create processes from scratch. There are some easily obtained checklists that can be reviewed and tailored to fit an employer's specific needs. There are also Excel spreadsheet-based tools to document processes, hazards, control measures, and more. There is more information on this topic in Chapter 11 on the Prevention through Design concept.

Many organizations and workers have heard of or used the job hazard analysis/job safety analysis (JHA/JSA) approach for identifying job task hazards and control measures. We discussed (in Chapter 4) that the creation of a JHA in a particular JHA format is not mandatory by OSHA (although it can be in some contracts), but is an often-suggested approach to learning about hazards on the job. I am mentioning them again as JHAs can fit into your organization's scheme for hazard identification. They could possibly also fit into a scheme for pre-work inspections in some applications, depending on how an organization creates and uses JHAs or modified JHA processes.

If your organization does or will use JHAs, expect that your workers will have questions about how JHAs are supposed to be completed, by whom, when, how they are supposed to be used (e.g., are they "gospel" or just suggested approaches), if or how they are to be retained, shared with customers, and so on. In my experience they can be a good technique or tool to use, but only if used intentionally and as more than a pencil-whipping exercise. In any event, the point is that providing some framework for how to accomplish a JHA or JHA-like process, whether as a mandate or a guideline, can be helpful.

Incident Investigation

There is no requirement from federal OSHA for private employers to have an *across-the-board* incident investigation program. There are some standard-specific requirements if you are subject to OSHA's Process Safety Management Standard located at 29 CFR 1910.119. Some State Plans have investigation requirements within OSH standards, and states can have requirements within workers' compensation laws.

Even without a universal investigation mandate for private industry,[7] injuries and illnesses that exceed certain thresholds require that incident data be recorded on the OSHA 300 and OSHA 301 forms or state equivalents. Some of the information is such that at least a certain degree of investigation is required, such as for what the employee was doing before they were injured, body parts affected, and a narrative description of what happened.

State workers' compensation laws require varying levels of injury reporting. Many require the filing of a "first report of injury" (FROI), although the threshold for type of injury requiring reporting can vary. The type of information required when a FROI must be submitted includes employee details and things like a summary of what happened, what was being done, and other things.

While some employers don't put much time and energy into incident investigations – others put significant emphasis on them. That can sometimes depend upon the potential severity of incident outcomes. Often receiving special emphasis are incidents that had fatal and serious injury and illness potential.

Something to point out here is that investigations should be approached as opportunities for learning and not with the spirit of "gotcha" inquisitions. It is important to know who did what and what went wrong. However, it is also very important to understand the background and circumstances for decisions made or why employees approached their work a certain way. Organizations do need processes for accountability. However, sometimes an employer's requirements are written in ways that are impractical, infeasible, or impossible to implement. Those are either poorly written or don't account for how work needs to be done. Were there discussions between the organization and the employees who must perform work under various conditions and circumstances?

For additional context, investigation with the identification of causes for events and necessary corrective actions, is a required element of common consensus standards for occupational health and safety.

Know that incident investigation is often considered a fundamental element in a safety program, even though organizations should also have "other" processes in place for identifying and controlling hazards.

Goal Setting

OSHA does not explicitly require an organization to set OSH goals and objectives within its enforceable standards. It advocates them, however, within its recommended practices documents.[8, 9]

Even without a mandate to do so, many organizations do set OSH goals and objectives of some sort.

The terminology in this area can be confusing, perhaps in part due to the evolution of standards as well as the fact that some organizations use consensus standards of different types, which sometimes use different terms.

Terms that have been used include goals and objectives, but also targets. Definitions can vary between organizations. For practical purposes, "goals" usually refer to what an organization hopes to achieve at a higher level. "Objectives" are things that help an organization to achieve those goals and can include some quantitation at a higher level. "Targets" can be used to refer to specific measurables that help an organization to determine if it met its objectives.

However, since uses can vary, let's just focus on the concept of trying to achieve OSH improvement in an organization. Focus on what an organization can measure that would demonstrate progress.

That could include measuring things like the completion of hazard analyses (JHAs or other), the closure of identified safety hazards, the completion of required training compared to plan, progress toward the completion of safety improvement projects, and so on.

Safety improvement should not just be about decreasing injury and illness numbers[10] in a given month, quarter, or year. While there can be a place for tracking those if not used exclusively, improvement can be better measured by looking at a combination of injury/illness metrics together with the accomplishment of actions that support injury and illness minimization. We also need to remember that just because something "bad" hasn't happened yet in some work situation or process, it does not mean that the potential for something bad isn't sitting unrecognized.

For frame of reference, voluntary consensus standards for OSH management and the OSHA Voluntary Protection Program require setting objectives, or goals and objectives. There is also a consensus standard that directly talks about safety metrics, ANSI/ASSP Z16.1.[11]

Formal Written Training Programs

On the Usefulness of Training Programs

Yet another program an organization may want but that is not required in an across-the-board way (although required by several different topical standards) is that of a "training program."

Your organization may have a requirement under one or more standards to have a training program that is responsive to the requirements of discrete standards. Sometimes a standard will require training as part of a higher-level program and spell out what it requires of that training. This applies, for example, to lockout/tagout/tryout (energy control) training and respiratory protection training. In these cases, the standard doesn't describe the training requirement directly as a need for a "training program," but for practical purposes, the standards describe many requirements for the training. It is more than a matter of having seat-of-the-pants training. At other times, a standard will reference a training program directly. This can include, for example, the powered industrial truck standard and the lead standard.

Then again, you may well have processes and activities taking place for which there is no OSHA standard specifically for your processes, but for which training is desired by your organization.

At the end of the day, one may be mincing words here. The bottom line is that OSHA has a lot of training requirements, whether OSHA describes those requirements as needing to have a program with required topics or to otherwise provide training on particular topics. OSHA's publication 2254–09R[12] can be a helpful reference, but be aware that it does not include newer training requirements established after it was last published in 2015.

You may well want to develop a training program that covers all of the organization's training needs, beyond the requirements of discrete OSHA standards. Without a written approach of some sort, it can become very difficult to be sure all requirements are met.

Your organization's training program (plan/whatever you want to call it), can not only specify training topics, but also when the training is required ("new hire," "initial," and "period" or "refresher" buckets) and how it will be delivered. It can be helpful for tracking training completion. Your organization could require different processes for different kinds of training needs, such as for formal classroom training conducted in house versus contracted training versus on-the-job training.

A comprehensive approach can be useful for the purpose of describing how worker understanding and competence may be or will be required to be measured. Reports that can be created from a tracking system can also be useful for communication with responsible managers or supervisors for planning, budgeting, and compliance tracking.

Some organizations can benefit from having a "learning management system" or LMS. For some, it could be great just to have a well-developed spreadsheet. It depends. There are some excellent consensus standards on the topic of training. Consider checking out ANSI/ASSP Z490.1 and A10.2 standards regarding safety, health, and environmental training. The A10.2 is specific to construction and demolition. You may also be interested in the ANSI/ASSP Z490.2 standard, which is about accepted practices in e-learning for OSH and environmental topics.

Training for Supervisors/Managers

Some employers have a special set of training courses that they provide to supervisors and managers with OSH-related duties that may be new to them. It is not uncommon for people to be promoted rapidly into supervisory positions due to workforce/labor considerations, before understanding that in their new supervisory roles, there is generally a higher level of responsibility than for the general worker. This can be true, for example, with respect to an employer's knowledge of hazards, enforcement of rules, and other matters.

Also, supervisors play an incredibly important role in making safety "happen." It's important that they grasp and embrace the basics and take responsibility for safety. Hopefully, they demonstrate a concern for worker safety in their capacity as leaders.

There are "regulatory" training needs, "company policy/procedure" training needs, and then there are just good practice/good communication and "safety leadership"-type training needs. It can make sense to have a training program that defines the employer's intent.

Some "safety leadership" training topics can include:

- Supervisor roles and responsibilities pertaining to management of the supervisor/employee relationships
- What it means to show "leadership" in OSH and environmental matters
- Management expectations for supervisors and employees
- A primer on basic safety concepts like hazard identification and corrective action processes
- Pre-shift and other routine safety inspections; supervisor participation
- A review of the employer's safety framework, e.g., the value of safety to the organization, the framework for how it manages safety, and requirements contained in programs, plans, procedures, etc.
- Management of OSH and environmental issues; support for corrective action programs
- Recognition of ergonomic risk factors and basic design guidelines
- The supervisor's role in incident investigation
- Obligations the supervisor needs to be aware of, which are built into the safety program
- Specific site items of high importance, goals, and objectives,
- The potential implications of regulatory noncompliance
- Processes for getting answers to safety questions and how to handle areas of ambiguity

In summary, there are many potential training requirements that employers need to be aware of and to provide for regulatory compliance. Some OSHA standards require topic-specific training programs.

There are many additional things an organization may want to provide training for, which pertain to how an organization otherwise wants things to run.

Safety Committees

Many organizations form safety committees that meet on a periodic basis. Federal OSHA does not require private employers to have them. However, 15 of the states with OSHA State Plans do require them, at least if the employer meets certain criteria such as when the company has more than a certain number of employees or has high injury rates over time, as measured by an insurance industry metric called "EMR" (experience modification rating).

If you have operations in a State Plan state, then you should check to see if one is required, and under what circumstances.

There are many different opinions about what a "safety committee" should look like. Everything from holding meaningless sessions that accomplish nothing except possibly eating pizza, to those that are well designed and make a meaningful difference to the organization.

If you are mandated to have a safety committee by a State Plan, then sometimes there are also requirements to describe its function, or to create certain documents to reflect safety committee activities and other things. Again, you will need to check on what your state requires.

Regardless of whether your state requires a safety committee, there can be good reasons for having one. While this can depend on the number of employees, some of the most effective committees I have seen spell out (in writing) things such as:

- Who will chair the committee? Will that be determined by job title or something else? Will chairmanship rotate every year?
- How will the organization seek its members?
- Does the organization intend to obtain representatives from each department?
- What will the intended function of the committee be?

  From what I have seen, some organizations do not optimize the use of their safety committees. It can be helpful to communicate what the intended purpose is through a safety committee charter, or something similar.

Is the purpose to obtain input from the plant or organization, as communicated from the broader organization to the safety committee members? Is it to actively participate in special safety improvement projects? Is it to participate in investigations or special purpose safety assessments, etc.?

- Training that members might receive.
- The role of the safety committee in hazard identification, risk assessment, and prioritization of issues.
- How long committee members could remain on the committee. Smaller organizations may have a more difficult time generating interest and/or making staff available to participate, based on staffing levels. Larger organizations could have more than enough candidates.
- How the organization will address participation if there are multiple shifts. Different shifts can have different issues.

  Safety committees can be a great thing for communication and involvement, or they can be a nightmare. Increase your odds of success by setting some parameters.

Documentation

If your organization is inspected by OSHA, its management can expect to be asked for particular documentation and evidence of requirements having been accomplished. Your organization could potentially be doing great things for safety but not get "credit" for some of it if it is unable to specifically produce what OSHA is asking for.

Therefore, you might want to make it clear in your written program what processes require documentation or records.

Inspectors follow inspection protocol described in OSHA's Field Operations Manual (FOM), Directive Number CPL 02-00-163, or an equally effective state equivalent in State Plan states. You can easily find it online and review it.

While inspectors may wind up asking for a variety of things based on what they see, some things that you can routinely count on an inspector asking for (and so you might as well have a plan to be able to share these) include the following:

- OSHA injury log (OSHA 300) and 300 A summary
- Your written safety programs/plans
- Hazard communication program training records
- Personal protective equipment hazard assessment

Other types of records likely to be asked for depend upon industry and the nature of worker exposure to hazards, but could include for example, respiratory protection training, lock/tag/try (energy control) training, forklift training, and others.

Risk Analyses

Some organizations may have a need for not only hazard identification but also a process to determine which hazards are most important to address. Having a process for that can be quite valuable.

While mandated for process safety management (PSM) regulated employers,[13] many other types of employers can also benefit from having "something" that can help it make sense of its hazards, what is acceptable, what should get paid attention to first, and so on.

From a practical standpoint, if your organization identifies 100 hazards, how will it know what to pay attention to?

Enter the concept of estimating the level of risk that can stem from these identified hazards, either by themselves or in groups.

Organizations with a need and the resources to do so can often and relatively easily take their hazard identification information and estimate risk levels. Broadly speaking, that is done by pairing identified hazards with corresponding likelihood of injury and potential severity of outcomes. It depends upon the complexity of an organization's processes. Details are beyond the scope of this book, because many options and levels of depth are possible. However, you can easily find simple examples online. As a caveat, I would say that these estimates are just that – estimates – and should be used as a tool within that context.

For frame of reference, estimating levels of risk can help your organization to manage its planning and budgeting actions. Also, the concept is well accepted by many employers across industries. It is advocated by many individuals who work professionally in OSH. The thought is that the use of risk concepts helps organizations to better integrate safety considerations into their bigger decision-making processes. It helps to connect safety to the bottom line of the business.

Contractor and Vendor Vetting and Coordination

Organizations that use contractors and vendors may wish to have a process to provide some level of assurance that when workers in these categories visit your sites, they will operate their processes and/or conduct their work safely with respect to their own employees as well as to work safely with respect to potential hazards from "your" facility or environment. The flip side is also true: It is important that the contractor's/vendor's activities will not impact the safety of employees in your organization.

It can be especially important to vet those organizations providing safety-critical services. That could include, for example, contractors hired to perform confined space entry rescue services, which is required by OSHA standards as well as related consensus standards.

The value in having a vetting process can vary based on your organization's size and resources, amount of contracting needed, type of activity to be conducted, inherent risk, and other factors. Bear in mind, however, that if people are on your premises or work location, you have obligations under OSHA to communicate certain things depending on industry, including energy control programs and in terms of safety data sheets (SDSs). If the work activities your organization manages constitute a multi-employer worksite, then your organization has the potential to be cited under various circumstances depending upon work relationships. Possible types of relationships can include being a controlling employer, a creating employer, an exposing employer, or a correcting employer. These terms, as well as how an OSHA inspector would look at relative obligations, are described in OSHA's multi-employer worksite citation policy.[14]

One other consideration is that when someone is "not" your employee but injured on your premises or due to your work activities, then third-party lawsuits might be filed against your employer. Seek appropriate legal counsel for specifics, which can vary between states. My point is only that third party lawsuits can (when a party is found liable) result in high dollar value judgments.

Other Programs to Consider

Change Management

Unless you are at a PSM-regulated facility (covered by 29 CFR 1910.119), there is no explicit OSHA requirement for an organization to have a written change management program.

However, having a change management program can help an organization to fulfill its obligations under the General Duty Clause.

I think that having a change management process of some sort can be more important than ever. Organizations are going through change at an extremely rapid pace. They've always gone through change in some form, but the advent of practical application of artificial intelligence will result in even more rapid changes to include some that could be accidentally detrimental to safety and yet difficult to detect.

AI (artificial intelligence) is starting to be used as a great tool for those managing OSH programs. It is concurrently resulting in major and rapid changes in manufacturing and construction processes. Improvements from an operational standpoint, such as for processing, batch processes, manufacturing technology, estimating, project management, and cost controls, while seeking increased operational efficiencies, could easily result in unintended negative consequences for OSH management. It will be increasingly important to have "something" in place to stay on top of how AI-driven changes, as well as other changes, can impact safety. Of course, people managing safety can also look for ways to integrate safety requirements and considerations into AI-driven software and other operational improvements to help ensure that potential gaps are addressed.

Warehouse Safety

If you have or will have warehousing activities in your organization, including high-volume retail applications, be sure to include those processes in your OSH program. Historically, some organizations have paid a lot of attention to the parts of the business where things are made or assembled, but paid less attention to the warehousing side of things. I am referring to where raw materials may be stored until they are moved elsewhere for processing, and to where finished goods might be placed or moved to, then packaged for shipment. Out of sight, out of mind?

However, processes that take place in warehouses can have significant risks and certainly need to be included in the scope of an organization's OSH plans. Don't make it hard for the warehousing folks to figure out how they fit into the safety plan. Warehousing safety processes should be integrated with the balance of the organization's safety management strategy and plans, or systems.

OSHA has stated that some types of serious safety and health hazards within warehousing operations include those related to:

- Struck-by
- Caught in-between

- Slips, trips, and falls
- Blocked aisles
- Means of egress
- Heat hazards
- Ergonomic hazards
- Powered industrial vehicles and other material-handling equipment

Other types of issues to pay attention to include meeting requirements related to:

- Fire prevention/protection
- Electrical codes (e.g., improper extension cord use, damage to wiring, listing and labeling requirements)
- Energy control programs (lock/tag/try)
- Conveyor guarding (from moving packages from one area to another)
- Ventilation (related to battery charging or other processes that require it)
- Respiratory protection (related to packaging or spills)

It's not that these types of hazards can't exist elsewhere in industry. It's more a matter of saying, don't forget that these can exist in warehouse operations. It's certainly important to ensure that you have written programs that pertain to those on the list that require one under OSHA. However, also consider how your organization may choose to identify and control hazards related to things for which OSHA does *not* require a written program but still poses a recognized hazard.

Another challenge in providing a safe environment and processes in warehouses is that warehouses are often leased. Sometimes safety issues can arise in those leased facilities that pertain to the property or its infrastructure, and which require coordination with and correction through the owner. If you are moving into a location that is leased, consider in advance what you may need to do to resolve OSH issues that could arise and which require physical modification or correction but that would require action by the facility's owner. You may need to interact with an owner or owner's representative to determine which organization will be responsible for what types of actions. For example, there could be OSH risks to your employees due to condensation from building processes that cause drips and create puddles on floors, resulting in slip or forklift operation–related hazards. If the underlying cause is equipment or processes owned by the property owner, then your organization would want a process to report such things and to coordinate corrective action by the owner. That could be either a practical matter or something that is required under the terms of a contract.

Know that whether you are operating in a company-owned or a leased facility, your organization has responsibility for its employees as described in OSHA regulations.

Another consideration is that both production and warehousing may utilize temporary employees. The organization's OSH program should make clear who is responsible for what and ensure an adequate and appropriate OSH program.

OSHA has requirements that pertain to your organization's responsibility to communicate certain things to its contractors (e.g., regarding energy control program requirements and hazard communication program matters), as well as to the provision of site-specific and process-specific training of workers under your organization's direction.

Safety-related consideration for contractors and temporary workers under an organization's control is a fundamental requirement of modern consensus standards for OSH management systems. OSH management systems are discussed in more detail in Chapter 9.

Stop Work Authority (SWA)

The concept of stop work authority is viewed and applied differently by different organizations. However, the phrase generally refers to an approach to safety where if an employee sees something that is unsafe or is being done unsafely, or if they otherwise have a concern about what they are being asked to do, they are empowered to stop, ask questions, and help ensure that work is performed safely. This is sometimes also extended to an organization's contractors and is also sometimes required by contracts. The phrase stop work authority can be unsettling to some; without understanding the intent, a lot can be read into it. That being the case, some organizations choose to have a process to apply the concept of SWA, but call it something different, e.g. "safe work intervention," "work pause," or similar.

There is no explicit OSHA requirement for employers to have a written SWA program. However, many employers do, and it is a much-discussed concept. The application of an SWA approach in safety programs is consistent with an employer's obligations under the OSHA General Duty Clause to provide a safe workplace, and an employee's right to refuse unsafe work that poses an imminent hazard of serious injury or death and that meets certain criteria.[15]

While there is no OSHA requirement to have a written program about SWA overall, it should be pointed out that the OSHA crane standard at 29 CFR 1926.1418 states that an operator must have the authority to stop work when there is "a concern as to safety" until a qualified person has assessed the matter and determined that "safety has been assured." Also, California's OSHA program has a process safety management standard for refineries that addresses the right to refuse unsafe work that applies more broadly than federal OSHA's approach.

Some organizations may view the concept of SWA as something that leaves the door wide open for labor disruptions. That is not the way the right to refuse work is described by OSHA. OSHA expects the employee to have taken certain reasonable actions and for certain conditions to exist before SWA is exercised.

While not law, as an example of support for the concept in industry, stop work concepts are described in the consensus standard ANSI/ASSP Z10.0–2019.[16]

Many employers apply the SWA concept more liberally than what is described in OSHA. They view it as something that promotes a positive safety culture and emphasizes the value that the organization places on safe work. It also provides the organizations and workers some flexibility to be able to think and plan through unfamiliar circumstances, changes in conditions, as well as processes and controls that are not working as anticipated.

If you decide to incorporate stop work authority into your programs, consider describing how the organization would like that process to work. This can help prevent unexpected delays and process ambiguity. For example, consider defining as part of your SWA process:

- The organization's policy regarding stopping work and the value of the SWA process
- What kinds of circumstances and conditions the SWA policy (or procedure) applies to
- Who the SWA policy and processes apply to (e.g., only direct hire employees, direct hire employees plus contractors, volunteers)
- The method for stopping work: necessary steps to follow, what to do regarding site or area control, who to notify, the prevention of other potential unintended consequences
- The process for investigation of the concern(s) that were the basis for stopping work

  (For example, describe who [which job position] will be responsible for the overall process of assessing and estimating risk, and which roles will participate.)
- Next steps and corrective actions: how the organization will determine and specify next steps to correct identified issues, how those actions will be taken
- The resumption of work
- How the organization will communicate to the immediate team and extended staff/organization about the stopped work and corrective actions

Industrial Hygiene/Occupational Hygiene Considerations

It can be easy for many people developing a safety program for the first time to think about physical safety concerns. Most of us have probably heard about ladder safety, the hazards of slipping and tripping, forklift accidents, and the person who sliced their fingers off in a saw and was then told to become a safety director. It is of course very appropriate to develop safety programs to address the range of physical safety concerns, including those discussed in Chapter 4.

However, what is less well known and less frequently thought of is the potential for employees to be exposed to and harmed by things like hazardous materials (e.g., toxic dust, fumes, mists, gases, vapors, fibers, and biological agents), the absorption of toxic materials through the skin, various forms of radiation, and noise. Many may also tend to ignore the potential risks associated with vibration, heavy lifting, static and awkward postures, and highly repetitive tasks.

Those types of potential hazards are often categorized as industrial hygiene (aka occupational hygiene) issues. The field of industrial hygiene (or occupational and environmental health and safety) is defined in different ways but is described by the American Industrial Hygiene Association (AIHA) as "the science of anticipating, recognizing, evaluating, controlling, and confirming protection from hazardous workplace conditions that may cause worker injury or illness."[17] Some types of potential exposures have been summarized in a nice way for easy use by the AIHA, as summarized in Chapter 4.[18]

It can be hard for an organization to know if worker exposures exceed regulatory limits (where they exist) or other guidelines. OSHA sets time-based limits for acceptable concentrations of substances in air in the workplace through various types of "permissible exposure levels" (PELs). They are intended to help prevent adverse health effects related to employee exposures. For a more limited set of materials and in a different context, the EPA has also set some limits under the TSCA.

Over time I've received questions about the acceptability of noise levels, dust, and powders coming off processes, metals from welding and plating processes, exposures to solvents, and other things. Correlated with that, there are questions about whether ventilation is adequate, if hearing protection is mandatory for certain types of work or areas, and if respiratory protection is needed to protect workers from toxic materials.

Those are great questions, and you often won't know the answer unless a competent person evaluates the exposures and compares them to established limits and guidelines. Although some people joke about it, our noses are not good enough, and it is possible to exceed limits and not realize it without sampling. Once again, it depends.

There are several materials that are hard to be sure you are protecting employees from, unless you have a competent person conduct occupational exposure assessments. That can include air sampling and other forms of assessment. Employers have a general obligation to ensure that employees' exposures are below the PELs, and OSHA mandates air sampling for some types of materials. There are several substance-specific standards for which OSHA requires an exposure assessment strategy or exposure control plan, e.g., for lead, asbestos, formaldehyde, coke oven emissions, respirable crystalline silica, and others. And air sampling is mandated for some processes, such as when welding on stainless steel, and for entering what OSHA refers to as permit-required confined spaces, where the atmosphere may be or could become hazardous. See 29 CFR 1910.146 and/or 29 CFR 1926 Subpart AA for OSHA's specific requirements.

From a regulatory perspective, as a simplification, if an OSHA inspector shows up to one of your organization's work locations on a given day, assesses the exposure of an employee to a chemical substance on that day and finds it to be over an OSHA limit, then the regulator may want to issue your organization a citation along with a penalty.

Obviously, there can be a lot of factors that might go into an OSHA Compliance Safety and Health Officer's (CSHO) evaluation, and that need to be considered. Depending on the material, the potential is there. On the other hand, the employee may work another 249 days that year or more, and the range of exposures over those other days could be significantly lower (or higher) than what the inspector obtained. You may not really know how to characterize the exposures to that worker and people performing similar work without more air sampling. The good news is that a few more samples and the application of industrial hygiene (also sometimes referred to as occupational hygiene) statistics by a competent occupational hygienist can give you a much, much better understanding of what is likely to be the true range of exposures ("exposure profile").

So what is my point?

My point is that, depending upon the type of work your organization performs, you may need to ensure that the organization conducts exposure assessments. Your organization is responsible for ensuring that OSHA's requirements are met, which will sometimes (depending upon what processes and materials your organization has) need to include an assessment of potential exposures and risks.

However, OSHA does not have an across-the-board requirement that organizations have a written macro-program describing their industrial hygiene commitments and plans.

Therefore, it may be helpful to mention something in your written program about industrial hygiene to help keep potential industrial hygiene–type exposures on your organization's radar. It is the employer's obligation to ensure that exposures are below regulatory limits. Depending on the substances your employees work with, you may need to build in specific exposure assessment requirements. For other cases, you may want to require a process that requires a competent person to assess your employees' potential exposures and to conduct or re-conduct exposure assessments as appropriate.

The Lone, or Remote, Worker

One other consideration for you is that people who work by themselves can have some unique risk factors. If your organization has people working by themselves, it makes sense to review their situations to help ensure that adequate protection is in place. After all, whether people are working by themselves or not, your organization has a responsibility for them under OSHA for occupational safety and health matters. And, in fact, there is the potential for things to occur that you might not ordinarily consider. Does OSHA require a written lone worker program? No. Does it make sense to be sure that your organization includes them in your programs to include hazards that can be unique? Of course.

Many organizations can have employees who, at times, may work by themselves and in circumstances where they may be considered a "lone worker." There is no comprehensive OSHA or NIOSH definition of the term, but it is usually used to refer to a person working by themselves and working without what might be normal organizational support, such as direct supervision or co-workers on a particular shift. It does not matter if the work is conducted during the day or at night. It can also include cases where an employee works around people but away from co-workers.

Lone workers can be found across industries. That can include industries and areas of work, including: retail, construction, manufacturing, utilities, medical care facilities, oil and gas, transportation, delivery, technical consultants, service technicians, hospitality, real estate, prisons, and family and protective services.

There are a few very specific (vertical) OSHA regulations that essentially prohibit lone work,[19] but none that are universal. There is no horizontal OSHA regulation that requires employers to have written programs to address lone work. However, the absence of thinking about special risks and taking extra precautions has resulted in worker deaths.

An employer's General Duty Clause obligations are clear. Practically speaking, one way to help ensure this is to have some form of hazard identification and possibly other controls in place for lone workers.

The risks to lone workers can vary. People working by themselves can be more at risk of:

- Non-violence-related injuries
- Violence-related injuries
- Personal health episodes

This can be due to the lack of ability of others to know there is a problem, modifications to the way people may choose to do work when not assisted by others, the lack of or decreased communication, more difficulty moving materials or equipment, decreased ability to see and operate critical controls and monitors, a decreased ability to respond to unusual emergency situations with equipment or processes, sudden releases of hazardous materials creating unsafe environments, weather conditions, or other things.

Approaches to managing lone worker risk, when intentionally addressed (yes, I am suggesting there can be gaps in industry), include, among other things:

- Establishing lone work policies and expectations
- Making clear what can and can't be done/prohibiting certain types of work/tasks (e.g., work from heights)
- Requiring pre-task work hazard identification/risk assessments
- Having special requirements for work in unoccupied buildings

- Mandating worker monitoring via "wearables" that can detect sudden acceleration and prolonged lack of movement
- Using fleet telematics
- Reviewing first aid plans to ensure they are adequate
- Providing extra communication between dispatchers and the remote worker initially and at remote project completion
- Ensuring periodic visits by security personnel and others
- Providing panic buttons/panic notification means

It does not need to take much effort to do *something* in the lone worker arena. Whether there are OSHA requirements or not, it makes sense to determine if your organization has risks in this area that it may wish to address. There are some excellent guidelines and resources available for your consideration, both from US-based organizations and internationally.[20]

Generational Differences and the Management of Safety

There have been many articles about generational differences in society as well as in the workplace, especially as it related to employer interactions with employees during the Covid-19 pandemic.

Curiosity seems to be significant regarding whether generational differences can or should impact the way organizations manage safety. There is recurring popular discussion in the media about what different generations value and how those values and perhaps circumstances impact how they view and approach work. One stream of discussion is about how the generations have differing expectations for the workplace and of themselves as workers with respect to work and what they are looking for out of employer-employee relationships.

Yes, there are some things that can be considered when developing your program, and which the organization can keep in mind as it implements its requirements. How you implement those can depend on your workforce and what aspect of safety management we are talking about. Here are some thoughts.

First – no surprise – there are some differences between generations that can exist. I say *can* exist because traits that might be most predominant in one generation can still be found in other generations. You've probably seen examples of this through an "old, young person," or those grandparents who keep up with the trends and concerns of much younger generations. With that said, common differences between generations can exist based on things like values, beliefs, work ethics, skills, and preferences.

The generations that are discussed most often, with corresponding birth years that are commonly referenced, are:

- The "traditionalists" (born 1928–1945); also referred to as the "Silent Generation"
- Baby boomers (born 1946–1964)

- Gen X (born 1965–1980)
- Millennials (Gen Y) (born 1981–1996)
- Gen Z (internet G, or iGen) (born 1997–2012)

You can easily find articles about generational traits online and in various books and other publications. From that, you can read some generalizations about traits that are most common within a group. There are both similarities and differences that can be identified. Often-cited examples of generational characteristics or traits (that are not always true for all people within a generation) include:

- Baby boomers: They can be idealists. They respect authority, are hard workers, and motivated by perks. Some started working before OSHA ever existed.
- Gen X: They can be in a saving and building-up mode. They are independent and feel like they've paid their dues. They are advocates for work/life balance who will potentially change jobs if there is not a fit. They could have issues reporting to people who are younger than them.
- Millennials (Gen Y): This generation is thought of as getting a trophy for "everything" and showered with praise growing up. They grew up during increasing digitalization. They embody a shift to valuing skills over degrees, with a desire for recognition for talents. They are purpose-led.
- Gen Z (internet G, or iGen): They place less value on hierarchy. They are visual learners based on how they grew up or are growing up, place a higher emphasis on "work vs. life," and focus more on mental health. They are very accepting of virtual work. As the youngest generation partially in the workforce, they are less experienced, and more likely to approach work plans from the standpoint of trying to figure out what they need to do to earn what they want.

Values that appear to be common between generations, include things like family, trustworthy leaders, a need for feedback, and opportunities to learn and develop.

The differences between generations can be a bit more complicated. My intent is not to turn this into a book about the details of generational differences, but instead to mention that there can be generational differences to consider when developing your program.

Some would suggest that organizations can modify various expectations to account for those generational considerations. I think that can be accomplished if done with caution, and how an organization does that can depend on many factors. In my opinion, the exact application depends on the context.

Keep in mind the following:

- The presence or absence of traits in a particular generation can vary.
- There can be as much difference between individuals within a generation as between individuals from different generations.

- "Business" impacts "safety," and vice versa.
- Local culture can be a strong factor and overwhelm generational considerations.
- Unions can strongly influence what OSH attitudes, beliefs, values, and implementation look like.
- We need to avoid absolute typecasting (which can be not only untrue but also insulting).
- We need to try to manage without our own generational blinders.

Obviously, some generations are rising as a percentage of the workforce and taking over leadership roles, as others are becoming a smaller percentage and giving way to younger leaders.

You may know that the Centers for Disease Control (CDC)/National Institute for Occupational Safety and Health (NIOSH) had a program referred to as the "Future of Work Initiative." This initiative identified issues that could impact the workplace, work, and workforce of the future, and established a research agenda in 2021.[21] A couple of the focus areas included virtual work and *mental health*. OSH challenges were listed, including the need for updated standards, the complexity of *human-system interaction*, and others. Additionally, wellness concepts were central to the agenda, and considered to be a basic human need.

The CDC/NIOSH Initiative identified issues impacting worker safety that exist and will not go away as important factors in safety outcomes. You can think about factors mentioned in the agenda when planning what to include in, and how to run, your OSH program.

How can these differences and considerations play out? How might these things be incorporated into the way your organization runs safety?

As hyperbole, let's imagine that a baby boomer, Gen X'er, millennial and Gen Z'er walked into a room to receive safety training one day. (No, not a bar, a training room.) The boomer might ask if there is a printed manual for the materials to be discussed. The Gen X'er might talk about how they just figure things out for themselves and ask if they really need to be there. The millennial might ask if there is something they can have to post on their LinkedIn profile showing their accomplishment when they are finished, and the Gen Z'er might want to know why they can't just figure things out as they go, working off their phone apps.

From a more serious standpoint, you may have people from all working generations together at once, whether for training, group hazard identification projects, safety meetings, or just present in work processes. Interests and aptitudes can be all over the map. You would not want to optimize for one generation so as to effectively cause the others to tune out.

An organization might, in part, seek to provide different options and means for interaction and learning. It could blend styles and elements. It could also develop processes to recognizing strengths and possible weaknesses of the workforce, strive to engage people from where they are "at," and facilitate a learning environment.

Your efforts could also include mental health and wellness concepts that can be helpful across generations. Is there an OSHA requirement for that? No. Might it make sense to consider workplace issues that can affect mental viewpoints, communication, respect, and working to create an environment that people are likely to want to work in? I think the basic answer is "yes"; but what that looks like in implementation may vary.

Here are a few things that you may wish to consider:

- Help millennials and Gen Z'ers recognize their potential and develop OSH skills.
- Listen. Millennials and Gen Z'ers often wonder, "Why are they not listening?"
- Take advantage of the vast experiences and wisdom of baby boomers.
- Try different communication methods to appeal to different preferences, e.g., digital vs. meetings vs. paper, desktop vs. cell phone.
- Consider how aptitude and preferences for use of technology can vary.
- See if different training methods can be used. Think about what you plan to provide in terms of formal vs. informal presentations and settings, lectures vs. self-directed learning, and microlearning.
- Consider that there are some very good safety-related apps that can be used. That includes for fall protection, crane signals, heat safety tools, and others. However, bear in mind that some people can believe that "any app is a good app," which is definitely not true. Consider reviewing and approving which apps and software the organization may endorse or approve for use.
- Success in the safety arena may look different to different generations. Provide balance and context in goal-setting processes and communication.
- Find and achieve common ground. It might be necessary to give in one area and gain in another. For example, a decision could be made to work on matters that will pertain to safety culture, rather than on a different safety matter that is important but that not all see eye to eye on. (Note: This is just an example. The big picture and context need to be considered in order to make the best decision.)
- Blend generations and facilitate dynamics that provide cross-generational learning. Implement processes that can foster understanding and communication that can in turn help avoid generation-based conflict. As an example, some companies have reverse-mentoring programs!
- Remember the potential for different physical capabilities. Design for flexibility where possible.

An organization *could* choose to be formal about how they consider generational differences. If so, it could consider, for example, an element of its OSH programs, and review how each generation may have unique things to consider and what an organization might do to leverage a generational trait.

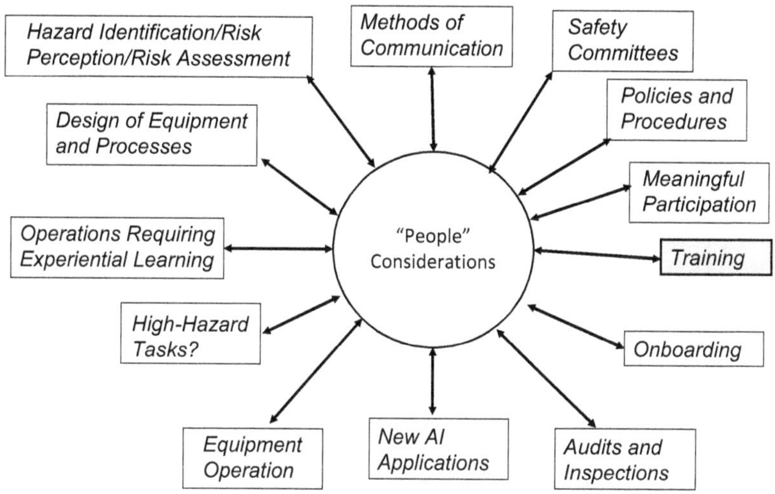

Figure 5.1 Sample Aspects of Safety Programs with People Considerations

As an example, let's say that an organization identified a number of people-related factors in its OSH program. The idea was to try to map out where the organization has processes that could be perceived differently by employees from different generations. It may generate something like Figure 5.1.

Going through that exercise, you can see that the organization identified a number of program areas that it may want to evaluate. That includes "methods of communication," "meaningful participation," "training," and several other things. Let's say the organization wanted to look specifically at the area of training, against generational considerations.

This fictitious analysis could look something like Figure 5.2.

You can see by looking at Figure 5.2 that the organization listed out things it wanted to consider that it thought might be important to each generation, and then assessed how and what it was doing to address each of those considerations within its training program.

Working through a process like this is of course not an OSHA requirement. I am also not suggesting that it is feasible for organizations to address every consideration it might identify. However, working through a process like that may help the organization to see opportunities to engage its workers in the most effective ways.

In summary, there are no absolutes. Consider that there can be differences not only based on generation but other factors as well. Seek to understand rather than stereotype. Learn from what people know and tie that to their unique skill sets where possible.

However – as a final thought in this section – shouldn't we want to do these things for our workforce *anyway*, accounting for some degree of variation in "people," regardless of generation?

Safety Element	Generation	Things to Consider	How Are We Doing That?
Training	Traditionalists	(Currently, none in our workplace)	(n/a)
	Baby Boomers	Structured, formal, expert-led	Opportunities for formal coursework in X, Y, Z
		Face to face	Coaching from senior leaders in Program Leadership 101
	Gen X	Self-paced learning and problem-solving	Training course 123 provides mix of traditional and digital methods
			Training course 456 provides practical examples that require some engagement but also self-study
	Millennials	Interactive, hands-on, collaborative, self-directed	Training courses X, Y, Z developed with gamification
			- microlearning for ABC - personalized learning paths
	Gen Z	Adaptable, quick to learn. Bite sized content, microlearning	Virtual reality, simulations
			Peer to peer interactions

And so on, for Meaningful Participation, Hazard Identification, Safety Inspections, etc.

Figure 5.2 Sample Analysis, Training vs. Generational Considerations

Consider Program Integration

Just a short word here on the topic of program integration, since earlier text might have created some natural questions and thoughts. The earlier text may be begging the question about how programs might, could, or should tie together.

Organizations that integrate their safety processes/procedures into the rest of the business and describe how the pieces are related and will be managed wind up being more effective. Providing some form of framework for how the pieces are intended to connect results in processes that are more aligned with the balance of the business.

A simple compilation of safety procedures, if developed piecemeal and addressing only one aspect of safety per document, may be *necessary* but not always *sufficient* for effective safety management.

Wrapping It Up

In summary, an organization's programs can help it to spell out the value it places on safety and health and make expectations clear. They can provide the organization's approach to where it (directionally) wants to go, describe the processes it will use to get there, and establish road signs and checks to help make sure it is moving in the right direction.

As for other aspects of business, organizations should be purposeful in their approach to occupational safety and health.

What is the alternative?

> "Would you tell me, please, which way I ought to go from here?"
> "That depends a good deal on where you want to get to," said the Cat.

"I don't much care where–" said Alice.
"Then it doesn't matter which way you go," said the Cat.
"–so long as I get somewhere," Alice added as an explanation.
"Oh, you're sure to do that," said the Cat, "if you only walk long enough."
Lewis Carroll, *Alice's Adventures in Wonderland*

Notes

1 In this context, organizations that are classified as employers and subject to OSHA, as mentioned in Chapter 1.
2 This book is focused on OSHA's requirements and organizational practices in an OSHA-regulated environment, or OSHA-approved State Plan equivalents. Many of the concepts in this book can apply across industries, such as for injury causes and management considerations (e.g., the unintended release of hazardous energy is the unintended release of hazardous energy, exposures to toxic materials can cause harm to the body regardless of regulator, and good communication is good communication). However, different regulators can have very specific requirements that differ from OSHA's, and liability frameworks can be very different. Be sure to consult the regulations appropriate to your organization.
3 Mandatory conditions for citing under the General Duty Clause include existence of a hazard, recognition of a hazard, risk of death or serious harm, and feasible abatement. Note the "recognition of hazard" does not have to mean that the employer was aware of the hazard. For example, recognition could be through internal records, the industry, or if a reasonable person would have recognized the hazard.
4 These tables were adapted from summaries published by the state of Tennessee's Occupational Safety and Health Administration (TOSHA). I updated information related to Hazard Communication Program (29 CFR 1910.1200) requirements. It might seem confusing that TOSHA provided a summary like this, since Tennessee operates an approved State Plan. However, it adopts many of the federal OSHA requirements as relevant to this discussion. Tables 5.1 and 5.2 are based just on the federal OSHA regulations. The TOSHA website from which this was derived bore the caveat: "TOSHA believes this information to be accurate and delivers it as a community service. As such, it cannot apply to every specific fact or situation; nor is it a substitute for any provisions of 29 CFR Part 1910 of the Occupational Safety and Health Standards as adopted by the Tennessee Department of Labor and Workforce Development or of the Occupational Safety and Health Rules of the Tennessee Department of Labor and Workforce Development."
5 It is up to the you to conduct your own research as to when OSHA or State Plans require written documents and what their content must be (as appropriate based on regulator).
6 California OSHA requires that employers identify the person or persons responsible for implementing an employer's injury and illness prevention program, in Title 8, California Code of Regulations, Section 3203.
7 Federal (non-private) employers do have investigation requirements contained in 29 CFR 1960.29.
8 Occupational Safety and Health Administration. (2016, October). Recommended practices for safety and health programs in construction (OSHA 3886). U.S. Department of Labor.
9 Occupational Safety and Health Administration. (2016, October). Recommended practices for safety and health programs (OSHA 3885). U.S. Department of Labor.
10 Many persons working in OSH are concerned about placing too great an emphasis on counting things after the fact (injuries and illnesses) as opposed to doing things to help prevent injuries and illnesses to begin with. Also, placing an emphasis on injury cases or rates can result in underreporting and detract from a culture that places an emphasis on learning. Note that when I refer to "learning" I am referring not only to learning from

what has gone "wrong" and being able to take appropriate action but learning from other things as well.

11 American Society of Safety Professionals. (2022). *Safety and Health Metrics and Performance Measures: ANSI/ASSP Z16.1–2022*. ASSP.

12 Occupational Safety and Health Administration. (2015). Training requirements in OSHA standards (OSHA Publication No. 2254–09R). U.S. Department of Labor.

13 Applicable to only some employers with hazardous materials of certain types exceeding threshold quantities, the OSHA Process Safety Management (PSM) Standard located at 29 CFR 1910.119 requires process hazard analysis (PHA). PHA processes involve hazard identification and evaluation of risk. Several methodologies are available. The purpose for the standard is to prevent or minimize the consequences of catastrophic releases of toxic, reactive, flammable, or explosive chemicals (see the regulation). Consult the standard for all its details.

14 See OSHA Directive Number CPL 02-00-124.

15 See: https://www.osha.gov/workers/right-to-refuse.

16 American Society of Safety Professionals. (2019). *Occupational Health and Safety Management Systems (ANSI/ASSP Z10.0–2019)* (pp. 14–15). American National Standards Institute.

17 American Industrial Hygiene Association. (n.d.). *About AIHA*. Retrieved May 25, 2025, from https://www.aiha.org/about-aiha.

18 https://www.aiha.org/get-involved/volunteer-groups/volunteer-committees-bodies-of-work/industrial-hygiene-hazard-identification-and-exposure-risk-assessment-by-market-segment.

19 This refers to the regulations that cover workers entering atmospheres that can be immediately dangerous to life and/or health (IDLH), work in what OSHA refers to as "confined spaces" (there is a special OSHA definition for that), work involving hazardous waste and emergency response, and work while near energized parts (for tree trimming and electric power generation, transmission, and distribution). It is also required for logging operations that someone be within "visual or audible contact," and for hot work, since fire watches are required. California OSHA also has regulations that relate to lone work. California's Title 8 Section 3395 addresses heat illness prevention and contains requirements for reliable and regular communications, which must be increased in certain heat stress environments/cases. California's Workplace Violence Prevention Rule from 2024 considers lone or isolated work to pose a violence hazard and requires corresponding coverage in workplace violence prevention (WPV) plans. Note that the Mine Safety and Health Administration (MSHA) also requires the availability of others in regulations pertaining to rescue.

20 You may want to look at: OSHA 3148: Guidelines for Preventing Workplace Violence for Healthcare and Social Service Workers; OSHA 3153: Recommendations for Workplace Violence Prevention Programs in Late Night Retail Establishments; Various UK, Canadian, and Australian documents (e.g., https://www.ccohs.ca/oshanswers/hsprograms/workingalone.html, and Government of Western Australia, 2009, "Guidance Note: Working Alone," at https://www.worksafe.wa.gov.au/publications/guidance-note-working-alone).

21 Tamers, S, Pana-Cryan, R., Ruff, T., Streit, J., Flynn, M., Childress, A., Chang, C.C., Novicki, E., Ray, T., Fosbroke, D., and Geraci, C. (2021). *The NIOSH Future of Work Initiative Research Agenda*. U.S. Department of Health and Human Services, Centers for Disease Control and Prevention, National Institute for Occupational Safety and Health. DHHS (NIOSH) Publication No. 2022–105.

6 Thoughts on Program Prioritization and Rollout

OSHA Program Rollout

Many people responsible for implementing new safety programs have questions about *when* mandated written programs must be in place, considering that placement of new processes and hiring of workers may be phased. A common area of inquiry is whether or not there is a grace period after operations start, before which the Occupational Safety and Health Administration (OSHA) would really expect written programs and training to be in place.

New safety leaders have also asked if they should implement other, voluntary programs they'd like to have in place, in the same time frame as for mandatory programs.

Necessary and "best" approaches can vary a bit, depending upon the industry, if the organization is in the early development stages or is catching up, and location of work. New construction projects can be yet another matter. A few considerations are described below.

New General Industry Location Start-up

The requirements applicable to new work start-ups can vary depending upon location (determining who your OSH regulator is) and context.

From a federal OSHA standpoint, if your organization is an employer subject to the respective standards, then there is no generic grace period per se. However, there is a specific enforcement policy regarding paperwork and written program requirement violations, described in OSHA Directive Number CPL 02-00-111. This document describes how inspectors are to treat paperwork violations that did not impact worker safety. Consult the CPL to see how it might be used. Several State Plans have adopted similar approaches. The point is, however, that the requirements still exist; they just might be cited differently than for some other issues, under certain conditions.

However, if your workers are exposed to hazards and there is a violation of an OSHA or State Plan standard, or if the conditions are such that the General Duty Clause of the OSH Act, Section 5(A)(1) applies, then your organization can be issued a citation for violations from day one.

DOI: 10.4324/9781032704586-6

From an industrial standard standpoint, if a new operation will be subject to the federal OSHA regulations and standards, then your organization should plan ahead. Work with your organization's staff involved in new process/facility implementation to learn who will be hired when, relative to which processes or equipment or facilities will be installed or built when, and relative to when the organization plans to bring workers on board.

Project plans, information on phasing, and so on can be very helpful. Sometimes key managers may not adequately realize that not all OSHA requirements can be met at the snap of the fingers, and that it really is the organization's responsibility to be aware of applicable regulations and comply with them.

That said, it is critical to communicate effectively and be sure that the organization is appropriately aware of OSHA regulatory requirements (such as for postings and administrative matters), program development requirements (which cannot always be done "overnight"), and training requirements (which can sometimes require significant advance planning). The idea is to help provide ample time to obtain necessary resources and prepare programs and plans consistent with the organization's desired rollout.

Order of Priority

The order of priority for your program development really depends upon which regulations apply to your work activities that require written programs, and what hazards your organization otherwise needs to address.

When it comes to topical, specific OSHA standards, there must be an exposure to hazards in order for OSHA to issue a citation. Recall that employers have an obligation to provide a workplace free from recognized hazards that are causing or likely to cause death or serious physical harm.

Depending on your work circumstances you might have access to plans and diagrams, or otherwise have an idea of what new work areas can require and what they might look like. You may well have a sense for what could become or be a hazard that will need some form of control.

If there are items that you know will be predominant in the workplace, and those should be evaluated to determine how your organization will address them.[1] What can you see, or what can you anticipate seeing? Plan to have programs in place as appropriate to protect workers from those hazards. If you are inspected by OSHA, the inspector would be likely to see those hazards and ask about them.

Look at the OSHA and State Plan regulations and determine what processes your organization will or might have and what tasks workers might need to perform that would be subject to them. The Tables 5.1 and 5.2 in Chapter 5 could be a starting point for you but should not be all that you rely on.

Go through a process to determine your organization's hazards and possible hazards. You could find some workplace hazard checklists online and use them if it helps, based on the type of areas your organization's workers will conduct work in. Perhaps your organization has other similar facilities that you could learn from.

As you walk through your facility or review what will be built, what do you see? For example:

- If you know that there will be work at heights, work on top of machines, work from elevated work platforms, etc., review walking/working surface and fall protection requirements.

  Protecting people who work from heights is a high-priority area for OSHA.
- If your new operations will have machines with possible hazardous motion and potential worker exposures, this can require a combination of machine guarding and energy control programs; review those to implement what is required.

  I recall being asked to visit a very small facility after a lead maintenance employee suffered a partial amputation. He had assumed that hazardous motion in an area he couldn't see had stopped and placed his hands blindly into moving mechanical equipment. The employer became concerned that OSHA would be paying a visit, so program improvements were made. It would have been much less stressful, disruptive, and costly for the owner to have conducted an easy assessment and implemented an energy control program up-front.
- Will there be chemicals in use, as part of production processes, maintenance, cleaning, packaging, or shipping processes? This would require a hazard communication program that meets the requirements of 29 CFR 1910.1200.
- Will there be pits, sumps, tanks, process vessels that can be considered confined spaces as defined under 29 CFR 1910.146, the Permit Required Confined Space standard? The organization would need something to conduct an assessment to determine if there are or would be permit-required confined spaces subject to the standard, and plan accordingly.

At that point, you should have at least some information that will help you if you work in a state that requires employers to develop an injury and illness prevention plan (IIPP), accident prevention plan (APP), or similar. If so, your earlier review can help you develop your IIPP or APP, since those are intended to describe an organization's approach to key program requirements and activities. There is no federal OSHA requirement that is overarching and comparable.

As mentioned earlier in this book, State Plan requirements can be significantly different from federal OSHA requirements, including having standards that are on different topics to those promulgated by federal OSHA.

For example, the relative priorities for an employer in California, based upon Cal/OSHA requirements and depending upon their work processes, might be:

- Injury and illness prevention program
- Heat illness prevention plan

- Workplace violence prevention plan/program
- Energy control (lockout/tagout)
- Hazard communication
- Emergency action plan
- Respiratory protection
- Hearing conservation program
- Wildfire smoke training

Note that the workplace violence prevention plan and the wildfire smoke training would not be required under current federal OSHA regulations, although OSHA provides guidance information on both topics.

Beyond that, your organization can develop its other needed written programs based upon the regulatory requirements, other hazards it knows it needs to address, and things it knows it will need to have in place in order to make its core written programs work effectively.

New Construction Work

An organization might approach its requirements slightly differently for newly planned construction work subject to construction standards. See Chapter 5's list as well as OSHA and State Plan requirements. State Plan requirements for construction can be more rigorous than the federal rules. If your work will be in a State Plan state, be sure to see the specific content requirements. For example, Michigan requires construction employers to have an APP in place, whereas it does not require the same for general industry employers. As another example, California OSHA (Cal/OSHA) requires its construction employers to have an IIPP as well as other hazard-specific written programs, but then, additionally, to create what it refers to as a **Code of Safe Practices**. So, developing an IIPP or APP and related documents could become a first priority for a construction organization, where one is required. In this case, as for others, templates available through regulators can make that relatively simple from a typing standpoint. The issue will be that the organization will need to determine what it wants to describe for its hazard identification and other plan elements.

Order of Priority

However, regarding general program development priorities, construction employers are often staffed and deployed differently than for general industry. It can often be the case that a construction employer must submit programs, plans, procedures, and even other things (such as their process for completing job hazard analyses [JHAs] – not an OSHA requirement) as part of vetting processes to be eligible to bid for work. It may need to submit those same types of documents and preliminary JHAs or job safety analyses (JSAs) prior to being awarded work or at least prior to being able to start work. If a construction employer has been asked for these before, then it probably has a core set of required documents, although it may need to create a site-specific health and safety plan, including emergency procedures, and revise or create some form of hazard identification and analysis for new work. Therefore, new written program needs might be less of an issue for a construction employer, due to common contractor vetting and project award practices.

If a smaller construction organization has limited procedures but is taking on new markets and larger projects, especially for larger general contractors, then it may find itself suddenly needing a complete set of requirements documents. In that case, as part of vetting or eligibility procedures, it is likely that an entire program or set of procedures will be created and submitted all at once.

If not, it can make sense to prioritize program development based on planned work, a combination of regulatory requirements for written programs, other hazards it identifies, and other processes it essentially needs in order to make the written plans work.

OSHA Tools Online

You may want to know that OSHA has a very helpful set of tools available for your potential use that can essentially take you through a quick-start process for program development and rollout. These quick-start tools are available for general industry, construction industry, and healthcare. The starting point link is at https://www.osha.gov/complianceassistance/quickstarts.

These links and related information files can be useful tools. If you are developing a new program, these pages can be helpful, as they contain an excellent collection of information. The pages start off by listing steps that can be taken to work toward having a compliant OSH program, broken down into seven steps (eight for healthcare). Then, links for each of the steps take you to other pages that provide access to resources, such as for the standards themselves, fact sheets, booklets, sample programs (templates), and other things.

The starting point page for general industry (from the link above) reads as follows:

- Step 1: OSHA Requirements That Apply to Most General Industry Employers
- Step 2: OSHA Requirements That May Apply to Your Workplace
- Step 3: Survey Your Workplace for Additional Hazards
- Step 4: Develop a Comprehensive Jobsite Safety and Health Program
- Step 5: Train Your Employees
- Step 6: Recordkeeping, Reporting and Posting
- Step 7: Find Additional Compliance Assistance Information

While the collection of information is quite good, note that some of the text describes things that OSHA suggests but that are not required by regulation. For example, there is a step about developing a "Comprehensive Jobsite Safety and Health Program." That is not a federal OSHA program requirement.[2] You may want to pay extra attention to what the regulatory requirements are as opposed to something "recommended." This is not to say that the recommended items might not be good for many employers, but not all recommended items will be a good fit for all employers.

Some would suggest that your written programs *not* be developed as one single and combined file, but rather be developed distinctly and be made available separately. That can make your processes easier to update and manage over time, and make it easier to supply requested documents to compliance inspectors (if/when inspected and requested).

Over time, you will want to keep track of organizational changes as well as regulatory changes that may require you to modify your written programs.

A Training Note

Obviously once an organization's written programs are in place, or as they might be developed (consistent with the OSHA steps), there are training requirements that the organization needs to address. Where training is required, OSHA mandates that it be provided before the work that triggered the need. This includes temporary workers.[3]

Plan for employees having had appropriate and adequate training before conducting the work that requires it.

A couple of questions seem to naturally stem from this. One is how to handle training for workers that are brought in from, say, a union hall. A second is how to handle training requirements during a site start-up, since it might be that the staff were hired precisely because of their expertise, and since there may be very few employees hired during the earliest phases of a start-up.

The answer to the first question is that the hiring employer cannot simply assume that the workers have had all necessary training simply because they came "from the hall." Union training has a great reputation and may well be totally adequate for most task-based hazards, but the employer may wish to have a process to ensure the adequacy of required training and ensure that documentation is available if required by OSHA standards for the type of work in question. Additionally, the hiring organization will still have company-specific and site-specific training (as applicable) for the work that it plans to ask the workers to conduct.

Regarding the issue of being in the early hiring stages and having, for example, just one employee in a particular job category, the employer has an obligation to have a process to ensure that prior training was adequate and that documentation is available if required by an applicable standard. Otherwise, it needs to have a process to provide the required training. Can the employee be considered competent to recognize hazards associated with their work? Can the employee stop working if necessary to ensure safety and obtain additional information and training?

Voluntary Program Rollout

Depending on circumstances, some employers want to develop and roll out their OSHA-mandated written programs before they choose to roll out non-mandatory written programs. This is sometimes based on concerns about implementing too much at once, or about employee buy-in for programs not strictly required by OSHA.

From a conceptual point of view, some organizations are small enough such that they could implement OSHA-mandated written programs *only*, and it would be clear within the organization how those programs would need to be rolled out. However, for many others, it can be very helpful to have some supplemental information that describes how those discrete program requirements will be managed and to describe how some important OSHA requirements will be addressed even when OSHA does not require something in writing. An organization may want to be careful about how it describes other non-mandated internal requirements if the organization knows that they will not happen consistently or be sustainable.

With that said, an organization might choose to implement non-mandated requirements at some time after mandatory programs are rolled out. In that event, I would suggest that the organization roll out those other programs as it would for other programs involving change. As we all know, people are resistant to change. There are some special things to think about, especially when making significant changes. While implementing any program can require "selling," implementing the above and beyond type of safety programs can take even more convincing.

Just like for other types of organizational change, think in advance about pros and cons and potential resistance to change. Think about how the organization can identify supporters of the new processes and develop allies. Work to ensure there will be senior leaders or sponsors of change who can help make things happen. Consider describing additional programs (if they are not mandated by OSHA) in terms of financial and intangible value to the organization, and in terms best understood by your receiving audience.

Consider providing extra communication, phased rollouts/trials, opportunities for worker participation, and achieving small successes that can be described and shared through the organization as implementation proceeds.

Incorporation of OSH into Other Business Processes

I've seen many organizations develop procedures and processes addressing individual safety topics but then not have a plan for how the organization will implement them all. What can end up happening is that these pieces of a program become orphaned and not implemented well or are implemented without review, understanding, or context.

Employees then don't understand how these safety requirements relate to each other, nor the importance of the pieces. When the safety program is not incorporated into the way the organization conducts the rest of its business processes, then substantial safety considerations may not be included in the organization's planning, budgeting, internal checking, or performance checks. They may not align with the balance of the organization's goals and objectives.

How successful your new program will be is not all about how long your written program is or even how well it is written, necessarily. To a large extent, success is about your organization's culture and how well leadership adopts your OSH program's principles and provides ways to ensure that the right things happen – in real life.

On Implementation – What Works in Practice

Written programs can be just meaningless words on paper. What an organization writes is not always what it "does." Once programs are written, there is the task of determining how best to make a given program/plan/procedure's requirements become reality. There can be both art and science to that. One could have an extremely well-written program but not take appropriate care to roll it out effectively or to follow the program as intended. The program could basically be left on a shelf, and not do anyone any real good. The organization's leaders could fail to "apply" the program well or consistently with best management practices. This is depicted in Figure 6.1.

Figure 6.1 Organizational Processes Not Running Effectively

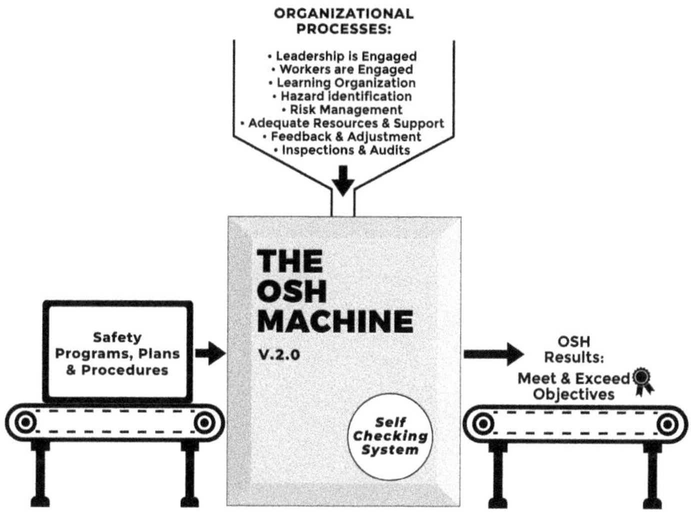

Figure 6.2 Organizational Processes Running Effectively

On the other hand, I've audited some organizations that had written OSH programs that were not ornate but were clear, simple – and used. The management teams were personally involved in safety leadership. They practiced what they preached and helped their organizations to understand the value of safety and of the organization's processes. Things were not always perfect, but the site leadership teams understood their safety processes and made sure they worked as intended. Processes included hazard identification, planning, and budgeting. A good number of employees "got into" their sites' safety initiatives. People were not afraid to communicate, and various activities helped to ensure that things were working as intended. *Their* safety programs ran more like Figure 6.2.

Organizations that take care to implement what they've written, and promote a culture that supports safety, can get the best results. It's not *only* about paperwork.

Interestingly, the reasons some organizations don't *apply* their written programs well can include the belief that they waste time and cost money. However, organizations that have ill-defined and piecemeal programs that are not implemented well may wind up spending the very money they are trying to save on safety fixes, patchwork, and injury expenses. They may be losing out on other opportunities as well, as summarized in Chapter 1.

Notes

1 You may have opportunities early in a new facility or new processes design phase to help "design out" potential hazards. That can save an organization a lot of grief later, and

possibly avoid deaths or serious injuries. This is often referred to as Prevention through Design (PtD), which is addressed in more detail in Chapter 11.

2 Although some State Plans require all or some employers to have components of a comprehensive program, as part of an IIPP or APP.

3 OSHA has published its Temporary Worker Initiative Bulletin No. 4, which describes host employer and staffing agency interface.

7 Who Should Do the Work?

Responsibility for Development and Implementation

If your organization is developing or improving a company safety program, it could be easy to ask yourself who should be "in charge" of the development and implementation efforts. After all, the written programs should be written well and reflect organizational specifics, describe actions that need to be accomplished, and need to be implemented effectively. There will also be many other responsibilities that someone will need to take care of, even if you don't lay them *all* out on paper (presuming that your organization does not yet have anyone dedicated to occupational safety and health [OSH] on staff, or available).

However, that is probably not the best question to ask first, nor the best way to conceptualize your safety program and approach! I will share suggestions for "who should do what," but consider the following.

First, remember that people in senior management and organizational leadership play a role, whether they realize it or not, in any safety management program or system. Leadership direction, buy-in, support, and communication from the top are known to be crucial. So, while an organization might have a focal person for the program, it is certainly not effective to think that "all things safety" can usually be handed off to one person. Usually, a team approach is needed to get the best results.

That said, an organization's senior-most manager is not usually involved in safety program development, except perhaps to spell out some things they feel are important to include or that they would like to see emphasized. They are critical in many other ways. Management leadership in OSH is considered to be a core element in the framework that it shares for effective safety and health programs.[1] However, it is still common for senior executives to delegate OSH development, implementation, and management responsibilities.

Second, the person who develops your program and the person who implements and manages the program could be different. Some people are great at writing and describing things but not implementing them, and some can *do* but not *write* programs well.

Safety programs and organizational requirements should not be developed in a vacuum. An organization's employees should be consulted and provide input to

DOI: 10.4324/9781032704586-7

program development, even though availability of resources can depend on the business phase, stages of hiring, and other factors. Coordination may also need to include people in corporate roles (if that exists) and other divisions or locations.

Asking who should take the lead role for program development and program leadership/management can be an important question but also may not be asking *all* the right questions and can potentially confuse your staff. This is because even though it is common for one person to take a lead role, safety processes, to be most successful, need buy-in and participation from team members with several different areas of focus.

Decisions about who the safety lead will be should be made carefully and in consideration of a variety of factors. Some variables are more important than others. Your selection should be based on a thoughtful evaluation.

Breaking It Down

Here are some things to consider when determining assignments. Organizations obviously may not use the same job titles referenced here, so apply the information conceptually.

Organizational Structure

What Part of the Organization Should Safety Staff Report Through?

Fundamentally, *where* the person who is taking the lead in safety for your organization reports to can make a difference in program effectiveness. How much of a difference it will make can depend on some company specifics.

In a smaller company with uncomplicated programs and lesser risks, the bulk of the safety leader's role may be about production, quality, maintenance, human resources, or other areas of specialty. Sometimes an individual can already be wearing several "hats" and then take on "safety" as well. If they are still able to accomplish what they need to accomplish for the safety program and the organization embraces safety well, then *where* that person reports in the organization may not be as important as whether or not the safety lead is respected and can get things done.

As safety becomes more critical to the organization, however, possibly due to things like company size, regulatory requirements, complexity, inherent risks, sheer workload, and safety culture, reporting structure can become more important.

A classic issue even in mature organizations is the perception that safety requirements detract from production or operational leadership goals, simply creating "more hurdles to jump" (although *you* know how safety can add value, as discussed in Chapter 1). With, especially, a weaker safety culture, the balance of the employees may be less likely to pay heed to safety-related procedures or requirements. Supervisory reporting can be especially important when dealing with that and other culture-related problems.

Experience has shown that it is important for the senior-most safety leader to have a "seat at the table" with the executive team and have "access" to the organization's senior-most executive. If the safety leader reports through a department that has no voice or no clout or has multiple reporting levels above them, and if the people in those roles don't know or support safety, then experience has shown that critical safety risk information can get filtered and not heard by appropriate decision-makers.

Can other reporting relationships work? Sometimes but not always. I have seen less than optimal reporting relationships work, but they have worked in part because of stellar skills and characteristics possessed by the safety leader (related to people and technical skills and personal drive), coupled with a positive safety culture and a leadership team that "gets" safety. However, that is not the norm. If you can create a safety program as you'd like it, I'd say find the person who has the skill set to communicate effectively up and down the reporting chain and who has aptitude, interest, and relevant safety related knowledge, and give very careful consideration to reporting relationships. Don't start off by impeding your effort.

How Many People Are Needed?

The Occupational Safety and Health Administration (OSHA) does not address this topic directly, in terms of numeric count. It's up to the organization to determine what it needs, except that some standards, depending on what the organization is doing, require competent and/or qualified persons (as OSHA defines them) to perform certain tasks.

Also, many organizations, depending upon what processes they have and regulations they are subject to, combine coverage for OSH responsibilities with environmental compliance responsibilities and other functions. The amount of time needed to support those other duties can be highly variable and make estimating the time needed to support OSH a bit harder to estimate.

There is no commonly accepted staffing level "norm" in the United States. The ratio of safety professionals to employees varies across industries. Some companies establish internal standards for safety staffing levels; these internal standards consider variables such as number of employees, degree of hazard as estimated by site type, scope of safety role, size of site, and other factors. You can easily look up some sample calculations online, but the criteria should be based on what it takes to fulfill responsibilities and company expectations – not on someone else's formula.

Staffing standards exist in some other countries. However, it is not straightforward to relate those standards to US employment situations because of differences in culture, regulations, occupational medical and insurance approaches, and other things.

Initial estimates of how much time will be needed to meet regulatory and other requirements are often driven by rough gap analyses as mentioned above, along with some benchmarking and comparison to others in your industry. The most important thing might be just to get started and develop a better "lay of the land." First estimates can be wrong and depend upon the competence and drive of the person taking the lead for safety, as well as what that lead person learns in their role and how well the organization does or does not support them in their role. As work progresses, the organization may realize it needs higher levels of safety competence and/or a higher number of available hours to meet its goals, or it may identify opportunities to do things more efficiently.

Another thing that can impact "how many" people are needed is how organizational leadership perceives the role of the safety leader or staff. For example, in some organizations, the safety leader functions as a technical leader and coach. While they lead, identify issues, plan, promote and improve OSH matters, and have safety-related responsibilities that they themselves address, they can also rely on a broader team of other managers, supervisors, and other interested people from other functional areas who also accept safety-related responsibilities. Those other people pitch in and help make sure that things happen, including for training, compliance, preventive maintenance, assistance with energy control programs, inspections, and leading through example. That's one scenario. In another scenario, the safety leader may not be working within a positive safety culture, or the organization itself may be running so leanly that no other leader feels like they have time to do anything except focus on their core area of expertise and keep their own head above water. And, in that case, and possibly detrimentally to promoting and encouraging safety ownership, other means of support or additional safety staff may be needed.[2] As above, however, the determination should start with gap analysis and reasonable planning.

Should We Consider Part-time Employees and/or Consultants?

Safety program implementers may feel they are just at the point of *maybe* needing help and needing to make safety leadership and employment decisions. They sometimes wonder if they can get by with developing their safety programs slowly and before making any direct-hire decisions. They would like to, instead, rely upon consultants or part-timers to get by until they feel their operations reach a certain level. Is that OK?

It's a great question, and a tough one. This may feel like a bit of a non-answer, but of course only your organization can determine what it really needs. If you know that critical risks exist or feel that there are large unknown areas that need exploration and development, then consider external help. Many organizations do use safety consultants to help conduct initial hazard and gap analyses and develop programs. However, you need to realize that OSHA's obligations are for employers and employees of a regulated organization. The use of a consultant does not change that. Organizations are held responsible for meeting safety obligations even during early start-up phases. Your organization is still subject to OSHA's inspection processes.[3]

Think about it this way, too: Just because an organization is not running full steam ahead, it does not mean that safety hazards are not present. Risk to the organization could be considerable because of a variety of factors. These can include the inexperience of new employees, starting up processes that are largely new and unfamiliar, differing attitudes toward safety and risk tolerance, high levels of equipment installation traffic, the breadth of processes that need to be paid attention to, change management considerations, and many other things.

Implementing safety programs can be like implementing many other organizational change projects. The organization needs to assess where it is "at," compared to what it is required to have and considering other possible leadership goals. Knowing the gaps in existing processes and future planned processes, the organization can develop integrated and prioritized plans for what it needs to develop, by when, by whom, and at what approximate costs. As for other rollouts, the organization can and should consider things like having process champions and steering groups, achieving wins in areas the organization can see and feel good about, consideration of company resources, internal and external pressures and market conditions, and other things.

Even with the help of an expert consultant, there are some things that internal employees know best and that internal employees need to make happen.

What Level of Employee Do We Need?

As a reminder from Chapter 5, I refer to the person who heads up or is assigned to take the lead role in running an organization's safety programs/processes as the "safety leader," regardless of what title an organization might assign. This is not meant to infer a particular seniority level or job category level.

That said and without intending to apply a particular societal *value* on job levels, organizations could experience difficulties kicking off a new OSH program with a person having minimal clout and authority. This can send the wrong message about the value an organization places on safety. It can, practically speaking, make things much tougher to implement due to political considerations and practical experience that can help one "get things done."

While some organizations may have rigidly defined job titles for OSH staff, job titles can really be all over the map. Some commonly used titles specific to the safety leader include Vice President, Safety Director, Safety Manager, Safety Supervisor, Safety Coordinator, and others. However, one company's Director can be another's Coordinator for a comparable level of responsibility.

Everyone has an ego and people tend to like nice job titles. Nicer titles might attract more interest. However, whether considering existing staff or new-hire (external) candidates for a lead role, prior titles can be deceiving. The organization should focus on a candidate's relevant experience, the size of organizations they've worked for, organizational authority and budget, areas of expertise, understanding of key OSH concepts, and how OSH can be integrated with other processes.

Know that there are a number of formal educational programs focusing on different aspects of occupational safety and health, exposure science, industrial hygiene, and environmental science at the bachelor's, master's, and PhD levels. Not all organizations will need to hire someone with a degree in the area, although there are some organizations that absolutely should. That can be a function of level of effort needed, complexity of the organization, technical or special know-how required, and other factors. There are some excellent degree programs with strongly relevant courses, and they can include great internship experiences. However, be aware that there are also programs that carry an OSH title but that may have little relevance to your primary need.

Some organizations have published hiring guides based on the intended hiring level of an OSH employee, with reference to expected skill sets, knowledge, job areas of focus, certifications, and other information.

Should We Pick Someone Who Already Works for Us, or Hire an External Candidate?

"It depends." Broadly speaking, there is no federal OSHA standard that dictates a particular educational background or other requirements for someone an employer chooses to fill a safety leader role.[4]

However, your company could be positioned in an industry and have processes that warrant having a safety leader with particular credentials that internal candidates may or may not have.

Also, your company could be in a position where it performs certain types of work for which OSHA (or Environmental Protection Agency [EPA] and OSHA) would require people serving in particular roles to have received special training (e.g., asbestos abatement), or to be "competent" or "qualified" to perform certain responsibilities (e.g., regarding fall protection).

Additionally, your company could be subject to non-regulatory contractual requirements and special conditions mandating that people filling certain safety roles have a particular education, experience, or certifications.

From a conceptual perspective, it can be a decision not so much about whether to assign someone internally versus hiring someone externally, as much as whether a particular individual has the requisite skill sets.

I have seen organizations start their safety programs by using internal candidates who have some prior background in OSH, or some training coupled with an interest in OSH. This can sometimes work because the candidate may already know the company's processes and culture and grow into the new role. If issues are basic enough, people in this category can often seek some help from a qualified third party or maybe work with the coaching of more experienced staff from other company locations. They could also work under an employee development plan. The suitability of that hiring approach really depends on how inherently risky the company's work processes may be, how complicated they are, how quickly conditions could change without the safety leader recognizing new hazards and responsibilities, and the likelihood of missing critical requirements.

I've also seen those attempts fail sometimes, because the assigned person took on additional safety responsibilities but was not relieved of other responsibilities, and/or they were otherwise set up for failure.

It can sometimes be abundantly obvious to an organization that it does not have a suitable candidate to become its safety leader. It is up to employers to be sure that they meet their OSHA obligations. External candidates are sometimes the best or only way to do that.

Can the Safety Leaders Do More than Just "Safety"?

Many organizations assign people to a safety leadership role, all the while expecting that person to do more than just safety work. It's really a matter of whether or not the organization's safety obligations are met. Employers should develop an understanding of their requirements and obligations and periodically assess whether they are being met. If regulatory obligations are not being met, then the employer should work to determine why. Job responsibilities may need to be adjusted, workloads may need to be shared differently, processes may need to be made more efficient, or staffing levels may need to change.

Should We Set Aside Time for This, or Can Someone Just Squeeze These Safety Things into the Day? How Many Hours per Week Are Needed?

In my experience, for people wearing multiple hats, it can be very difficult to do all the safety-related things required while being casual and simply "getting to them when they can." It can be very helpful to ensure that expectations, time obligations, and periodic but recurring needed actions be systematized to the extent possible and times be scheduled and set aside to be the most effective and maintain forward program direction.

If Someone Takes the "Safety Lead" Role as an Additional Duty, What Other Career Fields or Functional Areas Might Be Most Similar, and Where Might Synergies Exist?

Many OSH professionals can be excellent generalists as well as have highly refined skill sets in specialty areas.

That said, people working in other functional areas can sometimes already be performing tasks where parallels can be drawn to those needing to be performed for safety purposes. This is not to say that some areas of expertise in the safety world are not unique and distinct; there definitely are safety concepts and duties in that realm. However, there can be relationships. The relationships can depend on how specialized and complicated an organization's nature of work is. Where the strongest synergies lie between the different parts and processes of your organization will depend on a variety of factors, including the nature of your business, similar requirements that may be driven by other regulators, and other factors. Some non-exhaustive examples include:

- Maintenance staff could be well trained and suited to address physical facility requirements, energy control/lockout requirements, and machine guarding (although I've also known some departments to not understand those requirements and often ignore them).
- Human resources staff can align well with the *people*-related aspects of safety, such as for psychosocial safety, workplace violence prevention, allowing workers to meaningfully participate in safety processes, workers' compensation programs, and training and skills development.
- Environmental staff (if that function exists) will often have processes for work area inspections, investigations of non-conformances, along with corrective action tracking processes, and many other processes for environmental management systems that align well with requirements for (non-mandatory) occupational safety and health management systems. Examples include risk assessment, regulatory compliance, auditing, management review meetings, and continual improvement processes.
- Quality/process improvement staff will often have processes related to document control, risk assessment, investigation of causes of variation, requirements for change control, process quality, and improvement.

What About Having No Safety Leader or Manager at All? Does It Make Sense to Rely Upon the Team/Someone from "Across the Pond," or Elsewhere?

From time to time, I've encountered programs that are technically managed out of other countries and for which there is technically no local safety leader.

In my experience, those arrangements don't work well. Local staff tend to abdicate to other locations, and individual departments tend to not take initiative and otherwise presume different departments are addressing key issues. Regulatory frameworks from overseas parent locations can vary and may not fit well with local requirements. Apart from that, many times things seem to be "lost in translation."

Furthermore, some OSHA programs require that someone be appointed as a local contact for certain responsibilities. For example, respiratory protection programs are required to have a "program administrator" who must perform certain duties. Those duties appear to be at least extremely challenging if not impossible to fulfill from a long distance, based on the administrator's duties listed by OSHA.

Considerations for Selecting a Safety Leader

The first thing I'll point out is that, in my experience, people who are "voluntold" to take on a safety leadership role sometimes don't have a deep and abiding interest and correspondingly fail in their assignment. A little natural interest can go a long way. Further, employees do not always perceive that they have the time to do so, if the safety leadership role is presented as an "additional duty." The "ask" can be seen as a hindrance to other tasks, as opposed to an opportunity to help the organization.

With that out of the way, not everyone realizes that there are indeed many routes of entry into the OSH profession. This is in part due to the fact that OSH is a broad field. If you are hiring an OSH professional from outside your organization, you cannot assume that all people who call themselves safety professionals would be competent at what your organization most needs, or that all people working full time in the safety field have received the same fundamental training. Your organization should attempt to identify the education and types of experience most relevant to your industry and company. You may or may not select someone from outside your company, but know that guidelines for hiring and frameworks for competence are available, such as through the American Society of Safety Professionals (ASSP).

While there are indeed many routes of entry into safety roles, just because someone is performing certain safety tasks does not mean that they are the best candidate to lead an organization's overarching safety program. For example, let's say that someone is performing OSHA-required injury and illness recordkeeping and reporting, or that they are the go-to person for fall protection responsibilities. That person might be great at the task they are performing but may or may not be the best candidates to lead your safety program overall.

So, who and what type of person might be appropriate? Some organizations have the need and the resources to hire full-time, dedicated OSH or occupational safety, health and environmental professionals (as opposed to having someone with partitioned duties). They may be driven to do that based on regulatory considerations, board expectations, or customer obligations.

For that matter, senior level safety leadership positions may require significant credentials. This can be true, for example, for larger corporations, businesses in highly regulated industries, positions where the company can expect the leader to have higher public visibility, and for roles that involve helping the organization through complex technical issues and tricky regulatory issues. It can also be helpful to have those advanced backgrounds in order to best appreciate and manage evaluations involving, for example, significant engineering considerations or industrial/occupational hygiene evaluations.[5] Those kinds of positions can require advanced degrees and long, proven track records.

However, in other situations where higher-level credentials may not be needed, it is still helpful to fill the safety leader role with someone who has relevant knowledge, skill sets, and aptitudes. The "safety leader" role is not for everyone!

If you are bringing someone new into your organization specifically to take the safety leader role, then it makes sense to hire that person with some level of professional safety credentials.

If you are planning to take an existing employee and make them your safety leader, be careful. On the one hand, that approach certainly *can* and *has* worked in some organizations. On the other hand, I have heard from many people who entered safety roles with limited OSH experience that when they first agreed to take their safety-related roles, they had no idea there were so many components to occupational safety. Many reflected that they initially thought the role would be simple and involve not much more than common sense, only to realize later how much they did not know.

There are more things to think about and address in safety management than many people realize. The exact credentials and characteristics that an organization may want to look for in its safety leader can vary depending upon factors mentioned earlier.

Regardless, here are some things that can be good to look for when evaluating someone for a safety leadership role:

- Energy and caring; and excellent personal leadership in safety, a significant part of the "how" in how an organization makes its written programs work, can be a huge factor in safety success
- Attention to detail
- An ability and willingness to engage workers in learning about their jobs and hazards
- A basic understanding of the applicable federal OSHA or state-run OSHA framework
- An understanding of OSHA inspection processes and frequently cited standards
- Completion of various external safety training courses such as for one of the OSHA 30-hour[6] or OSHA Standards classes (OSHA 510, OSHA 511). and other special-topic classes (e.g., recordkeeping, fall protection, respiratory protection, machine guarding, energy control [lockout], etc., depending on company exposures)
- A willingness to continue training in technical and administrative safety concepts
- A knowledge of what causes occupational injuries and illnesses and common approaches to prevention
- An understanding of and comfort with common hazard identification and analysis techniques
- Adaptability
- An appreciation for and willingness to work with employees from a variety of backgrounds and with different learning styles and perceptions

- Demonstrated initiative toward making OSH-related improvements
- A willingness and ability to speak in front of groups of employees
- The ability to write clearly and succinctly
- The ability to communicate and discuss key safety risk information with management, which refers not only to the ability to package information in terms of risk but also the ability and willingness to clearly express information of a regulatory, company, or ethical nature that may need careful presentation due to pre-existing safety culture and prior perceptions

Professional Credentials

As mentioned earlier, there are professional degree programs in occupational safety and industrial hygiene/occupational hygiene. They are available at different levels and from organizations of highly variable reputations. Some programs are excellent, sometimes even with relevant internships mandated to receive a degree. I also know of some programs where the professors/instructors are incredibly involved and even go so far as to support attendance at professional association meetings and even as guests at standards-setting meetings. Some also have strong mentoring programs.

As for any degree program, it is still the case that programs can be of variable quality, areas of focus, and elective opportunities. Know that if you are reviewing resumes, just because someone has a degree in some aspect of OSH may *not* mean that the candidate is the best fit for what your organization needs; there may be other people looking for positions where coursework and emphasis were more aligned to your organizational need.

There are also professional credentialing bodies for those working in OSH. These include the Board of Certified Safety Professionals (BCSP) and the Board for Global EHS Credentialing (BCG). Certifications issued through the BCSP include the Certified Safety Professional (CSP) designation, the Associate Safety Professional (ASP), the Graduate Safety Practitioner (GSP), Occupational Hygiene and Safety Technician (OHST), the Construction Health and Safety Technician (CHST), and others. Certifications issued through the BCG include the Certified Industrial Hygienist (CIH), and others with environmental and product stewardship areas of focus. Credentials obtained through those organizations are meaningful and valuable.

There are other organizations that also offer meaningful credentials. However, there are some paper mill–style credentials that can be found on the market. When reviewing your applicants, you can and should check on the veracity and credibility of claimed credentials. I've seen many resumes over time that listed credentials that were misleading and/or meaningless.

Know that recent college/university graduates may not be initially eligible for some professional certifications. That can be because of certification eligibility parameters that require candidates to have a certain number of years of experience. However, they may still be great for what you need, depending on things like maturity, prior degrees, and/or experience. Therefore, look carefully at your hiring needs at present and as they may exist in the future. You could hire someone who will be what you need at the moment but also really grow with the organization over time.

Work Assignments Built into Written Programs/Plans/Procedures

Lastly, here are a few thoughts regarding the description of roles and responsibilities as contained in an organization's safety programs.

Except for an organization's OSH policy statement or vision statement (typically a very high-level document), it is standard for written program documents to specify the roles and responsibilities of different parts of the organization and/or different functional roles. Required action steps are also built into stepwise procedures or flow charts.

While I've tried to share ideas in a positive way, I do need to suggest a couple areas of caution regarding how you specify who will do what.

Those items are:

1. It is helpful to write your documents in a way that makes it clear how processes are supposed to work. If there is insufficient detail, then readers may not understand who is supposed to do what. When roles are not clearly called out, people may presume that some other group or individuals will take care of responsibilities, even when in fact "they themselves" were intended to be responsible for taking care of certain responsibilities or described actions. There is an old saying that "when everyone is responsible, no one is responsible." That saying is not meant to convey that we don't *all* have some degree of responsibility for safety. Rather, it is saying that if duties are described generically, over-broadly, and without an adequate degree of specificity, then human nature is often to assume that "someone else is taking care of that," or that with everyone being responsible, someone else will wind up taking care of the duties.

   My suggestion is, if you write something, make sure there is a reason for it and that it will make a difference to your organization.
2. Please don't write your programs in a way that says the safety lead will, practically speaking, do all work for anything that has the word "safety" in it. Once in a while I see programs written that way – with little or no reference to other functional areas. That is not a sustainable or effective way of running a safety program.

   The safety lead will certainly have many responsibilities across the span of safety programs. However, that person cannot do all things and be omnipresent. It would not only be impossible to do them all but also send the wrong message to the organization. Safety management should be accomplished as a team effort and with the involvement of many, for best

> success. For organizations with multiple departments, it is recommended that program documents describe and require the interaction of the groups for both support and expertise.

Also – and I've seen this in several organizations – if an organization does not describe how responsibilities are shared, it may not feel or be natural for leaders from other functional areas to pick up the flag. Being realistic, managers and supervisors hired to help an organization focus in other areas tend to be trained in and enjoy those other areas. They may have no background in "safety" whatsoever and may have little interest in it; that's not a fault but a function of academic programs and personal interests. So even though persons from across the organization need to take responsibility for their safety and for others, they may not know what to do. They may need to be encouraged and coached about how they relate to an integrated safety program, what the organization now expects, and how they can help.

Safety programs should not be written for the sake of checking a box or for a blind vision or to just be able to say you have a safety program. Rather, they should be written to describe expectations for how processes and interactions will work.

Notes

1 https://www.osha.gov/safety-management.

2 Many safety people have found themselves in roles at some point where safety culture needs improvement. This can be a very challenging situation. Easier said than done – the safety leader should act to influence, teach, and coach in a structured way.

3 OSHA's processes for inspections are described in the Field Operations Manual (FOM) Directive CPL 02-00-163. This covers topics including inspections, violations, penalties, and many other things. You can find it online.

4 There are, however, some specific training requirements for people performing certain types of work, or serving in certain functions or roles, such as when performing asbestos abatement. There are also expectations that persons serving as "competent persons" have an ability to identify hazards related to the hazards addressed by a specific standard, and the authority to get things done. OSHA also references "qualified persons" in several standards, with references to requirements related to things like education, training or experience, an ability to solve and resolve problems, and other things. Readers need to consult applicable OSHA standards to see the exact OSHA requirements. Even so, the requirements are related to specific standards, as opposed to describing the qualifications for an organization's overall safety leader. Of course, if the safety leader is also performing competent person or qualified person duties, then that leader needs to meet the OSHA expectations for a person acting in that capacity.

5 Industrial hygiene, or occupational hygiene, is described by the American Industrial Hygiene Association (AIHA) as, "scientists and engineers committed to protecting the health and safety of people in the workplace and the community" and states that "[t]he industrial hygienist is part of a broader family of professionals often referred to collectively as the practice of occupational and environmental health and safety." Practically speaking, this can include things like taking air samples and noise samples to compare worker

exposures to occupational exposure limits (OELs), evaluating local exhaust ventilation systems, and many other types of evaluations.

6 The OSHA 30-hour class (designed for workers with some safety responsibility and for supervisors) provides an overview of safety concepts such as hazard identification, avoidance, control and prevention, supervisory skills (relative to OSHA obligations), and other overview information. The 30-hour class does not make someone a safety professional, and it is far from a degree in occupational safety. However, it does contain a lot of very good information. It is available for general industry and construction regulations. There are a variety of OSHA-authorized trainers, both online and in person. In my experience, quality can be variable. As touched on previously, the OSHA 510 and 511 courses are also available, covering the construction industry and general industry regulations, respectively. The materials can be extremely helpful for persons who need to know regulatory framework as well as have an overview of the requirements of particular standards. However, taking a 30-hour, OSHA 510 or OSHA 511 class does not in and of itself provide all the information and skills necessary to serve as an organization's safety program leader.

8 Programs and Systems

What's the Difference?

Programs vs. Systems

> All things can be deadly to us, even the things made to serve us; as in nature walls can kill us, and stairs can kill us, if we do not walk circumspectly. … The least movement affects all nature; the entire sea changes because of a rock. … In each action we must look beyond the action at our past, present, and future state, and at others whom it affects, and see the relations of all those things. And then we shall be very cautious.
>
> Blaise Pascal (1670)[1]

Up to this point, we've reviewed some important and fundamental information, points that provided some context about programs and which should be helpful to most who are developing individual or multiple plans or programs. We discussed the many reasons for having safety and health programs, the significant benefits that can be obtained from them, what *drives* the need for them, common causes of injuries, the regulatory standards for which OSHA most frequently cites employers, common occupational safety and health (OSH) program elements and pointers, and even things to consider when specifying who in your program will *do* which pieces of work.

However, the information presented so far has been about what employers need to do or should do to address, mostly, focused types of hazards and requirements that OSHA standards specifically "call out." We've discussed how the Occupational Safety and Health Administration (OSHA) requires the employers it regulates to have written programs and plans to address hazards related to things like the machines, tools, and equipment in use, and chemical or toxic material–specific standards.

An example of a topic-specific OSHA requirement to have a written plan is the OSHA Process Safety Management (PSM) standard located at 29 CFR 1910.119. That requires an organization to have a PSM program – if it handles, stores, processes, or moves highly hazardous chemicals or processes flammable liquids or gases above certain thresholds. That standard references the hazard communication standard and the use of Safety Data Sheets (SDSs). However,

DOI: 10.4324/9781032704586-8

the central purpose of the PSM standard is the prevention of catastrophic events like explosions and toxic chemical releases that could stem from unsafe chemical interactions and processes.[2]

Those and other standards are important. Organizations need to know the OSHA requirements and make sure they are complying with them.

However, it can be helpful to think in terms of a bigger picture, and how different parts of an organization's activities form a system, whether we pay attention to that or not. Our work processes are dynamic, and what happens in one or more areas can affect one or more "other" areas, for better or for worse. With systems thinking, we try to be cognizant of the potential interactions between work processes, and monitor and adjust what we do based on those definite or possible interactions. Systems thinking is not new but something we can bear in mind when we develop our programs.

When an organization thinks about the larger organizational framework and superimposes checks and balances to make sure its broader set of processes is working well and in fact continually improving, it is applying systems thinking. It is acknowledging that the whole may not equal the sum of the safety program parts. When organizations very intentionally take systems factors into account, define the components of their safety program and documentation, and ensure the inclusion of some core safety management concepts, they are using a safety management *system* (SMS), or occupational health and safety management *system* (OHSMS). Some OHSMSs also require the use of a risk-based approach.

However, there is no OSHA standard that mandates employers to create that overarching "total program description" that describes its system for managing safety. No OSHA standard requires an organization to have a comprehensive description of how the plans, programs, procedures, and other pieces fit together or impact one another. Nothing says an employer must describe how it will manage other processes that are needed for effective management but that are not explicitly identified by OSHA as a requirement. (Is this overthinking things? For some, yes. For some, no.)

The closest that OSHA-based regulations come to requiring an OHSMS for organizations it regulates is when and where a state-run OSHA program requires an injury and illness prevention program (IIPP) or an accident prevention program (APP). Requirements related to IIPP and APP development were discussed in Chapters 5 and 6. The IIPP and APP requirements in question (at least at the time of writing this) don't require *as comprehensive* an approach to safety management as described in the pre-eminent OHSMS models (American National Standards Institute [ANSI]/American Society of Safety Professionals [ASSP] Z10.0[3] and International Organization for Standardization [ISO] 45001:2018, Occupational Health and Safety Management Systems[4]). They do, however, require more than just what is required in other discrete and focused-topic standards. So, it is important for each organization to be aware of the requirements of the states that it works within.

You may want to know that while neither federal nor State Plan standards require the organizations they regulate to have anything referred to as an OHSMS, the federal OSHA website nevertheless has guidance documents that encourage systems type processes. For example, OSHA's Recommended Practices for Safety and Health Programs located at https://www.osha.gov/safety-management (see the safety and health program "core elements") provides recommendations that are consistent with several of the requirements of the ANSI/ASSP Z10.0 and ISO 45001:2018 standards.

There is one other thing I should point out in this section. And that is that while the ANSI/ASSP Z10.0 and ISO 45001 documents are referred to as "standards," they are non-mandatory consensus standards, at least in the US. They are not incorporated by reference into the OSHA standards, and not legal requirements under OSHA.[5] (Consensus standards – what they are and how they fit into the US regulatory framework – were discussed in Chapter 2.)

Common Content in OHSMSs

So then, what is it that an OHSMS might have that typical topic-specific OSHA standard, or combination of OSHA-specific OSHA standards, wouldn't address or require?

That is a broad question and depends on which standards we are considering. There are some differences between the standards.

However, there are many similarities, and I would think about those. Overall, an OHSMS describes an organization's OSH policy and management intent, and the processes it will implement and maintain to have a viable, lasting safety management system. It contains requirements for making sure that there will be a way to keep tabs on key metrics and activities, and that there is feedback between processes. Regulatory compliance is intended but just a part of the bigger picture. The bigger picture is to make sure that the pieces of the puzzle work together well, know interactions between processes, and monitor them for adequacy. It is also to have activities in place (part of processes) for which there is no specific OSHA standard but that is still something it is expected to "make happen" under OSHA, such as providing a safe workplace.

Here are some examples of management system requirements contained in one or more of the popular OHSMSs. This is at a summary level; there are more details on some of the concepts later in the chapter. At a high level, in order to *conform to* what some OHSMSs require, an organization needs to, among other things:

1. *Evaluate context/content/forces:* Use what is likely already known about the business environment and make a conscious effort to think about how that or those environments can impact the safety of work. Where more information is needed, learn more about the internal and external forces

(such as the market, technology, etc.) that influence or control how the organization runs. Learn who "interested parties" are, what their interests and concerns may be, and consider how that can affect how work gets done safety. Use this information along with knowledge of the organization's products and services to determine what should be included within the management system (describe this in the OHSMS scope).

2. *Provide leadership and encourage participation:* Provide management leadership that truly supports and promotes a culture that is positive toward safety; a culture that will result in safe work processes and methods of communication and learning; one where leadership demonstrates interest and an intent to help work through roadblocks that inhibit safety; and one that promotes and encourages professional communication about safety concerns or improvement ideas, without fear of retribution or ridicule.

   Continuing with the concept of involvement, provide opportunities for meaningful worker participation – at all levels. That refers to involvement that is at more than the "I told them they could attend the committee meeting on their free time, but without pay" superficial level.
3. *Apply strategic and tactical planning:* Build safety into both strategic and tactical planning. For example, have processes to create larger and even multi-year objectives and targets, use hazard and risk information when developing plans and budgeting, and incorporate data from multiple other processes. Have a hazard identification and control process. Create plans based on identified gaps and successes and latch onto improvement opportunities.
4. *Ensure support:* Provide intentional systems to identify and provide the support needed to succeed with the organization's OSH goals and performance. This goes beyond lip service and includes categories for people, equipment and tool resources, education, training and competence, and document controls.
5. *Integrate processes:* Have integrated processes that include safety. For example, intertwine as appropriate the identification of issues and risk assessment into other business processes. Have processes for advanced design review, use of the hierarchy of controls,[6] and management of change. Be sure that even things that we might not think of as including safety incorporate safety matters, such as in procurement, contractor selection, and evaluation. Think about and include consideration of occupational health where it could be an issue in work (not just traditional physical safety). Meeting emergency preparedness requirements is also a common element.
6. *Use the right data, smartly:* Don't collect safety information just for the sake of building bigger data centers. Require processes that add value. Include processes to measure the *right* things. OHSMS's commonly require processes to monitor, measure, and assess performance of processes and of the system itself. They also require processes for incident investigation, the conduct of internal audits, identifying and taking necessary corrective actions, and, essentially, organizational learning.

7. *Hold routine and periodic management reviews:* The organization needs to be sure it periodically reviews the big picture for both planned processes and for the OHSMS itself. It is expected that what it learns from these reviews it takes and uses to further the OHSMS. That could include, for example, the modification of processes or procedures, incorporation in planning and budgeting cycles, and so on.

Again, there can be real differences in how any two similar organizations might apply a given OHSMS's requirements. However, the intent would be to implement the processes in a way that is consistent with what the organization *knows* to do and can learn it needs to do, and to have a system with resilience that can address planned, unplanned, and unanticipated factors that can affect safety.

Readers are advised to obtain and read any OHSMS they might be interested in. There are also guidance manuals or documents available related to some OHSMSs. These can provide helpful background information and discuss some of the newer theories about injury causes and prevention, as well as many (other) practical pointers. Some of these pointers are related to areas where an organization may have risk, but for which there is no explicit OSHA standard.

Should Organizations Use Them?

Should organizations develop and implement the use of an OHSMS, since that is not an OSHA requirement? Does that even make sense?

My answer is, "sometimes, but it depends." There are various factors that can affect the answer, such as the size of an organization, complexity of processes, supply chain environment, customer expectations, and … how detailed an OHSMS we might be talking about!

One way of thinking about an implemented OHSMS is that it describes an organization's approach to safety at a high level – the types of processes that it will have to both comply with regulations and to conform with other internal requirements that an organization feels are important to meet its OHS values and policy. An OHSMS in part provides a framework for how it intends its processes to work together and describes the elements in its system.

Certainly, for an organization to have some sort of list of its policies, procedures, instructions, forms, etc. that exist somewhere in its plans and sub-programs can be helpful. Practically speaking, managers and supervisors should know this information to be able to best communicate the organization's requirements and to administer them.

If your organization's managers don't know that your company has a safety program or safety procedures, or an OHSMS, well, that is usually not a good sign (speaking with a bit of sarcasm and based on real-world experiences). It's understandable that they wouldn't have all the safety procedures memorized and committed to heart, but if they don't know their organization has a program or where to find it, that is not an encouraging sign.

With that said, having a comprehensive management system is not appropriate or necessary for all organizations. While having one can make great sense for some organizations, it could be infeasible and of limited value for others – at least based on currently available OHSMS models.[7] Factors can include span of control and complexity of operations.

For example, smaller and medium-sized organizations in lower-hazard industries might have just one or a few managers, with each providing management and direction for different functional areas. There might be "no way on earth" that the managers and supervisors in those cases will have a need to use a comprehensive OHSMS approach. This type of organization *does* need to have regulation-compliant programs. However, creating a comprehensive system for the use of just one or very few people who supply the glue for the organization could be like talking to yourself or writing something about a grand strategy that only you will read. It might or might not make sense to put in that effort, unless you have customers or some sort of industry association that pushes you in that direction. Certain types of safety approaches and tools just make more sense in some organizations than in others.

On the other hand, while a formal OHSMS may not be a good fit for some organizations, there are others for which an OHSMS would be.

Where Can an OHSMS Be a Good Fit?

Some organizations "can" have a need for and get value by managing their safety processes with a systems-based approach. In general, there appears to be a growing recognition of the value of OHSMSs across the globe by countries, agencies, industries, and companies.

There are several reasons why. Sometimes it's because of the belief that an OHSMS provides a "North Star" for program direction, and a tool for use by both management and workers. It gives senior managers something to point to as a commitment that the organization has and that requires leaders to "calibrate to," in a sense. Some choose to implement an OHSMS because it demonstrates a higher-level commitment to the safety of workers and can send a message of caring to the workforce.

Others work in complex systems and value having an OHSMS because it should help identify critical business issues for risk management and provide continual improvement-based processes.

In other instances, organizations implement an OHSMS because their supply chain requires it to have an OHSMS, or to be certified to one, to be an approved vendor.

Yet another theme is that organizational leadership wants to set itself apart from competitors and enhance its reputation. They know that the disciplines and processes required by safety management help in other areas of a business as well. There are numerous case studies to this effect, including those described in an article by Joe Wolfsberger and me.[8] Some further details follow.

Where and How Can an OHSMS Make a Difference?

First, notice that implementing an OHSMS will require spending time, energy, and resources for implementation! As for almost anything, the more a person or an organization puts into something – presuming intelligent effort – the more they'll get out of it. The same can be true here.

With that said, here are a few high-level points where having an OHSMS can make a difference:

- Regulatory compliance
- Injury reduction and the minimization of risk
- Safety process improvement
- Improved business processes
- Organizational reputation
- Increased eligibility to win contracts

Having an OHSMS in place is to a significant extent about having *processes* in place. One of the typical process requirements of an OHSMS is to evaluate compliance (e.g., ISO 45001) or to have a policy that commits to compliance with laws and regulations (e.g., ANSI/ASSP Z10.0). An OHSMS can help a company comply with basic OSHA requirements, because of the basic expectations of an OHSMS and the dynamic processes expected.[9]

Some organizations have implemented an OHSMS for the purpose of injury reduction. I know some people who believe their company's implementation of an OHSMS was an important element in safety management but that it also improved other management processes of theirs. It can depend on what your starting point is. Incidentally, I am aware of various facilities over time that had less than stellar injury performance that were "encouraged" by a regulator to implement an OHSMS approach.

There seems to be a need for more formal studies that evaluate the impact of an OHSMS. It can be hard to know which program elements, or which safety initiatives, provide the most impact. For example, was it the OHSMS itself that made an impact, forces of regulatory pressure, or an organization's special safety emphasis program focused on some high injury rate area? Even so, there are some studies showing that application of an OHSMS can provide injury reduction value and better performance in key areas.

For example, one study looked at the value of implementing an OHSMS of a particular type in creating safe operations. The study was based on the Occupational Health and Safety Assessment Series 18001 standard (a British OHSMS standard that had been applied by organizations globally, and that was phased out as a result of the creation of ISO 45001). It evaluated proprietary performance data over a 21-year period, from groups providing registration to the standard, and the U.S. Bureau of Labor Statistics. The findings estimated that certification to the standard lead to a reduction in the total number of cases by 20% and that the cases involving job restrictions or job reassignments were reduced by 24%. It also indicated that certifying to this OHSMS – meaning that the employer met the requirements of the standard – was an indicator of "superior average safety performance."[10] There are other studies showing a positive performance correlation as well.[11]

Further, other studies have identified that implementing a management system approach did better regarding processes considered key to safety, such as the identification of risks, risk analysis, and corrective actions.[12]

There can be other business-related value (besides having safer processes for workers and minimizing direct and indirect injury costs). For example, there can be value in setting a few extra internal expectations for leadership, giving your workers a higher bar. Another example is the benefit of being able to tell your customers that you have an OHSMS in place, as opposed to just unconnected basic safety procedures. Saying "yes we maintain an injury log" is a lot different than saying, "we have a complete approach to keeping our people safe and on the job." For some businesses, being able to report having an OHSMS in place can make your organization eligible for contracts that it might not otherwise get. (Some supply chains even require reporting of safety management system–related metrics.)

All in all, I believe that certain types of organizations can benefit from the injury prevention and process improvement that can come from implementing an OHSMS. There usually are reasons for picking one type of OHSMS over another, primarily tied to the market an organization is working within and what it is trying to achieve. There are some good articles that you can find through the ASSP and elsewhere that compare elements and choice considerations.

Some organizations obtain certification to ISO 45001 through third-party registrars. Data available from the ISO's website indicate that there were 512,069 sites certified globally, and across many industries. That includes manufacturing, construction, and utilities.[13] The organization needs to be large enough, and have enough bandwidth and risk to gain the most from implementation.

The ANSI/ASSP Z10.0 standard is an excellent US-based standard. The basic standard is clearly written. There are also helpful background and guidance documents related to the ANSI/ASSP Z10.0 that complement it and can serve as an excellent primer in some basic safety management concepts.

OK, Then, So What's "Above and Beyond"?

There are a few concepts built into an OHSMS that go above the discrete safety programs or plans you are required by OSHA to have. This book is not intended to focus on OHSMSs, but I think it's important to highlight a few things for you.

Systems Thinking

First among those highlighted things are the benefits of the **"systems thinking"** approach. This refers to the intentional use of a coordinated, big-picture approach. It recognizes that various parts of an organization's processes influence each other in some way; sometimes less and sometimes more. It can be hard to make a change in one area that does not in some way affect another area, for better or worse. Processes are interrelated, whether organizations see that or not.

The approach also recognizes, and tries to some extent to account for, the fact that organizations operate within a certain *context*; there are internal and external forces that can impact the best-laid safety plans. Certainly, things like decreased margins, increased regulatory pressures, difficulties engaging workers beyond the limited parameters of a job description, decreased budgets, and changes in demand for products or services all have the potential to influence how an organization identifies hazards and manages risks. It would seem intuitive to some, but not everyone, that changes to the "forces" that affect any part of an organization also has the potential to affect safety processes and outcomes.

Why is this important and how is this different from just complying with basic OSHA requirements? Because OSHA does not require organizations to connect so many dots. Because the failure to acknowledge those links and account for them is tantamount to running your safety processes "by chance."

Not planning for how business processes can impact safety or otherwise not keeping an eye on forces that can impact your programs can create risk to your safety programs without the organization realizing it. And looking at things from the opposite side of the mirror, negative workplace safety outcomes and surprises can affect an organization's operations, e.g., daily operations, strategic planning, the availability of resources, including capital to be able to seize business opportunities, freedom from excessive third-party scrutiny, human resources considerations, including recruiting and employee retention, and public relations.

I know some organizations that had unexpected, unplanned-for negative safety-related events that threw their organizations upside down, in a way that likely could have been prevented if they had used a systems thinking approach. Processes can be unexpectedly shut down, with sudden unplanned costs that impact other opportunities, which can cause investors to grow nervous.

While not all bad things are predictable, organizations have an ability to influence the safety of operations utilizing more than "chance." They can improve their odds and add some resilience. So, why not look at things a bit more proactively, and intentionally address systems considerations as part of integrated routine business processes?

On the flip side, where systems thinking was used, organizations have prevented headaches and avoided costs. Some of those savings were achieved through advanced project management that included the consideration of safety requirements early in plant or new process design considerations, through the use of a multi-disciplinary team. They demonstrated the use of what some refer to as "Prevention through Design" (PtD) concepts. The PtD efforts resulted in complying with requirements more cheaply and effectively, working with the most realistic project timelines, designing processes that minimize injury potential, and even making it easier for facility staff to perform facility maintenance tasks.

Continual Improvement

Second are the benefits of the use of a Plan-Do-Check-Act (PDCA), **continual-improvement** system.

The ANSI/ASSP Z10 and ISO 45001 standards operate on the concept of applying a PDCA approach and describe organizational processes as dynamic and nonlinear. If a company chooses to conform to the requirements of one of these standards, it must apply processes with the intent to achieve *continual improvement.* Each standard's required processes are intended to be part of a PDCA framework.

The concept is that after an organization applies an OHSMS process, things don't just "stop." Rather, the organization will have systems in place to check if its plans, goals, and actions taken are improving outcomes. There will be periodic, routine processes in place to assess the business environment, plan action, take action, and check on performance and the performance of the changes. Are you on target or off target? Periodic small adjustments serve to make your business ever more efficient every time you accomplish portions of the cycle. Some organizations have continual improvement processes without having a full OHSMS in place, but having one in place will require that the organization build that approach into its safety processes.

OSHA's general industry and construction regulations don't require employers to have continual improvement processes. *But wouldn't your organization benefit from that?*

Risk Assessment

Third is the application of **risk assessment** concepts to safety programs. One of the most important and key elements of an OHSMS that goes beyond programs focused on compliance only is to have processes in place that not only identify hazards but also assess the level of risk associated with those hazards.

Risk assessment itself is not new and is used in many professional applications. However, risk concepts are not always applied to organizational safety programs and might be new to some.

OSHA documents themselves often refer to occupational safety and health *hazards* and requirements to control them. For example, a basic tool used by many to identify hazards and corresponding controls is the job *hazard* analysis. The word "hazard" in an occupational safety context refers to something that could cause injury or illness. However, just because something is a hazard, it does not mean that it poses a *risk.* When there is exposure to hazards, then the level of risk from those hazards can be estimated. There are various methods of doing that.

Organizations might know they should identify occupational safety hazards, but not, practically speaking, what to do with the information. Some organizations create long lists of hazards without having some sense of which might be more important than others, and which require action.

There are several methods that can be used to assess and describe risk. Being able to communicate in terms of risk can be an incredible communication tool for people working in safety roles. It can aid in the conveyance of important issues using terms that business leaders are likely to see when managing other aspects of business. Reducing uncertainties related to safety outcomes reduces uncertainty in the business itself. Performed competently and professionally, this can transform someone from being seen as, for example, an alarmist, to someone who has a grasp of business and conveys important issues in a level-headed way for further consideration.

In summary, the concept is to use risk assessment processes to evaluate risks and prioritize potential actions. This is not feasible for all organizations and situations. It can depend on many factors, such as scope of work, level of staffing, and in-house skills. However, the concept and practice do not have to be as complex or fear-inducing as some might make it out to be. The inclusion of safety risk considerations in how an organization runs its processes can be a great tool to keep priorities straight, develop action plans, and estimate the effects of control measures.[14]

Business Integration

A fourth, important concept is the **integration of OSH** with the organization's other normal processes.

My first introduction to one concept of the integration of safety into other management elements came somewhere in the 1996–1997 time frame while working for a company that was a contractor to the Department of Energy (DOE). Efforts from around that point in time wound up becoming what is now referred to as DOE's Integrated Safety Management (ISM) system. ISM includes concepts about the integration of environmental, health, and safety programs into work planning and execution, and is geared toward activity-level work.

Guiding principles are intended to provide a framework for an effective ISM system. ISM documents (published by the DOE) specify five core functions. Those core functions in and of themselves contain useful approaches to *part* of what an organization can do for safety management. They are consistent with core elements of other more recent safety management systems but focus on activity-level work and are not as broad as some others. DOE guidance documents also establish safety culture focus areas, which include leadership employee/worker engagement and organizational learning.

However, the meaning of *integration* of safety into other operations has been applied a bit differently in other settings and has changed over time. The DOE ISM approach addresses safety management in an integrated way but with an activity-level focus. The concept of integration of safety into something broader than activity-level work and as part of an organization's broader management framework and processes were mentioned in the 2005 and 2012 versions of the ANSI Z10 standard.

It was not until the 2019 version of Z10 that information about integration concepts was expanded. The 2019 version of the standard describes that integration in a broader way, especially regarding the impact of safety on the *business* (as opposed to the concept of primarily just including *safety* into the planning and execution). The guidance manual that accompanies the Z10.0 standard, ASSP GM Z10.100: 2024, Guidance and Implementation Manual for ANSI/ASSP Z10.0–2019, Occupational Health and Safety Management Systems, provides a chapter on this topic.

My point is that the "integration of safety" as a phrase is not new, but concepts for application have evolved. There are many organizations and businesses that see safety management as something meaning more than creating a few procedures that address a handful of physical hazards. They see it as an opportunity to create learning and increase system capabilities.

Figure 8.1 is one graphical representation of how business processes could interact with each other within one organization's OHSMS.

Here is what Figure 8.1 depicts:

- An organization has some type of core product(s) and/or services(s). This is represented by a central, dark, gray-colored wire.
- Surrounding the creation of core product(s)/service(s) are various functional processes, such as those for maintenance, engineering, and operations/production. Each of those functional processes works together with and is joined by occupational safety and health processes. These processes (maintenance/safety, engineering/safety, etc.) are represented by strands of two-shade wires that are wrapped around the core. For example, the thick-dashed black wire and plain white wire are wrapped around each other and represent the concept of engineering and safety processes working together in support of the common central theme (core) of products/services.
- There is a lighter-gray cover over the core and its surrounding wires. This represents a "condition of safety" that results from the underlying processes.
- The outermost layer simply represents the "total process" of the organization.

The overall point is that in an intentional OHSMS, "safety" can be tied to functional areas processes, and all the functions/processes in the system ultimately interact in support of the core product/services. Safety is integrated with, and a part of, functional processes, but also a condition of safety emerges for the whole. There can be safety outcomes that would not previously have been conceived or possible.

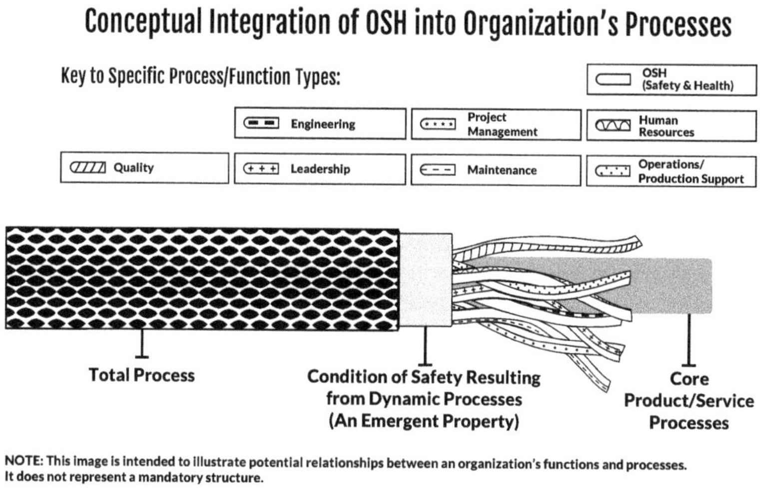

Figure 8.1 Illustration of How Processes Could Interact in an OHSMS

Evaluation of OHSMS "Fit" and Options

Upon hearing about the potential to implement an OHSMS, some business owners immediately think that it would be too costly to do so. The reality, however, could be that, if an OHSMS is right for your business and implemented well, the rewards will be well worth the effort. It could save money, increase compliance, decrease risks, and help protect you from liabilities.

Different models for OHSMSs are available for potential use. Some of these models are contained in voluntary consensus standards created by non-regulatory standards setting bodies. The basic elements of the various models are often similar, although with some slight differences. ISO 45001:2018 and ANSI/ASSP Z10.0–2019 were mentioned previously. Although they are not the only models that one could look at, they are the ones that I will share further information about here.

At their core, ISO 45001:2018 and ANSI/ASSP Z10.0–2019 are both focused on occupational safety and health and include requirements for core types of OSH processes. They both incorporate systems thinking and risk-based thinking. A simplistic description of the concept behind systems thinking is that one cannot usually make changes to one aspect of a business without impacting other functions in some way, and sometimes those interactions can affect safety for better or for worse. The concept behind risk-based thinking is to proactively understand an organization's risks, incorporate that information into management processes, and look at opportunities to reduce risk.

By virtue of their design, they acknowledge the interrelatedness of dynamic business processes and require actions that help account for that.

Both standards are designed to give you a solid foundation and system to allow the organization to monitor workplace performance metrics, and make changes as needed.

Differences between the standards can include some safety terms and the way topics are categorized, the number of numerical sections, the order of presentation and emphasis of topics, the depth of descriptions, and other things.

The ISO 45001 standard could be a strong choice, especially if your company conducts business internationally and has a supply chain that requires implementation of the ISO 45001 standard. Third-party registration/certification is available for the ISO 45001 standard, whereas there is no current option for third-party certification to the Z10.0. ISO 45001 is well known, and as an ISO management system standard has some content that is written in a manner like that of other ISO management system standards, e.g., ISO 9001 (for quality management systems) and ISO 14001 (for environmental management systems). There are similarities in language and there can be synergies between required processes. There are also several ISO 45001–related guidance documents that could work in a complementary manner. These include but are not limited to:

- ISO 45002, Guidelines for Implementation of ISO 45001
- ISO 45003: Psychological health and safety in the workplace – guidelines for implementation
- ISO 45004: Guidelines for performance evaluation of occupational health and safety management systems
- ISO 45006: Guidelines for preventing and managing infectious diseases in the workplace.

Others may be developed in the future.

The ANSI/ASSP Z10 standard can also be an excellent fit for some. It is written in an understandable way, with a substantial explanatory appendix. As mentioned earlier, there is also the related Guidance and Implementation Manual, and Guidance Manual: Keeping Your People Safe in Smaller Organizations.

Organizations can implement the Z10.0 standard incrementally based on need and other factors. When in conformance (meeting the requirements of the Z10.0 standard and of the organization's OHSMS), the organization should be able to easily respond to the supply chain and other interested party requests for information about its safety management system processes.

There are many similarities between the requirements of Z10.0 and ISO 45001, but some may find the Z10.0 easier to understand and use. Note that work done to conform to the Z10.0 standard should be helpful if the organization determines later that it would like to (or is told that it needs to) obtain third-party certification to ISO 45001.

Final Thoughts

Whether an OHSMS would be a good fit for your organization or not, I suggest that you do *not* think of your safety program as just a collection of individual and unrelated compliance-focused documents. At the very least, when structuring and managing your programs and making business decisions, consider how individual parts of your safety program can interact with other parts, both for better and for worse.

Keep in mind that whether you choose to use an OHSMS approach or not, the effectiveness of whatever you put in place will be impacted by more than what you place on paper. There are some great books available about systems considerations, learning from success, avoiding a blame-driven orientation, human and organizational performance (HOP), and many other relevant topics.

This book is intended to provide you with some basic and helpful information that you can use in the development of an OSH program. Not all organizations need an OHSMS. To the extent that I share information about OHSMSs, it is to provide you, the reader, with some information you can use when determining how you want to approach program development.

Notes

1 Pascal, B. (1670). *Pascal's Pensées*. https://www.gutenberg.org/files/18269/18269-h/18269-h.htm.

2 The OSHA PSM standard, where applicable, requires an extra level of skills, training, procedures, and processes. If your location is subject to PSM requirements, you may and probably already do know that, because of the nature of processes subject to PSM, chemical volumes, planning and permitting requirements, and other things. However, if you find that your organization is subject to the OSHA PSM regulations (and possibly also the Environmental Protection Agency's [EPA] Risk Management Planning [RMP] regulations under 40 CFR Part 68) and yet it was not previously aware of that, your organization will need to conduct extra research to determine appropriate next steps. This may be the case for any of a number of other OSHA requirements that require plans but for which site specifics can be highly variable and involve complex issues.

3 American National Standards Institute and American Society of Safety Professionals. (2019). *Occupational Health and Safety Management Systems* (ANSI/ASSP Standard No. Z10.0–2019). https://www.assp.org/standards/standards-topics/osh-management-z10.

4 International Organization for Standardization. (2018). *Occupational Health and Safety Management Systems – Requirements with Guidance for Use* (ISO Standard No. 45001:2018). https://www.iso.org/standard/63787.html.

5 While OSHA does not have a direct OHSMS requirement, note that a couple of US-based federal agencies require either an SMS or systems safety approach. The Federal Aviation Administration (FAA) requires an SMS for certain aviation-related entities. The Federal Railroad Administration (FRA) requires certain passenger rail operators to implement a systems safety program (SSP) that has a relationship to SMS concepts. Coverage and application can be complex; if your organization is subject to those requirements, then it should become fully aware of when and where they are applicable and which specific SMS elements are required, and implement them as required.

6 This refers to a preferred order in which *controls* are applied to keep workers safe. OSHA and the National Institute of Occupational Safety and Health (NIOSH – recently merged into the Administration for a Healthy America [AHA]) talk about this. The National Safety Council (NSC) does as well. The concept is that some ways of controlling safety hazards can be more effective than others. For example, elimination of a hazard is better than an engineering control, and an engineering control is considered better than the use of personal protective equipment (PPE). See https://www.cdc.gov/niosh/hierarchy-of-controls/about/index.html.

7 There are some guides for small and medium-sized organizations available through both the ISO and the ASSP. The guide published through the ASSP is available for free on its website, and is titled ASSP GM-Z10.101–2019: Guidance Manual: Keeping Your People Safe in Smaller Organizations.
8 Wolfsberger, J., and Zigulis, G. (2023, April). Integrating OSH into the business. *Professional Safety, 68* (4), 30–34.
9 However, it is the case that OHSMSs are intended to help organizations move away from a "compliance-only" focus and to apply a broader perspective.
10 Viswanathan, K., Johnson, M.S., and Toffel, M.W. (2024). Do safety management system standards indicate safer operations? Evidence from the OHSAS 18001 occupational health and safety standard. *Safety Science, 171*, 106383.
11 Mohammadfam, I., Kamalinia, M., Momeni, M., Golmohammadi, R., Hamidi, Y., and Soltanian, A. (2017). Evaluation of the quality of occupational health and safety management systems based on key performance indicators in certified organizations. *Safety and Health at Work, 8* (2), 156–161. https://doi.org/10.1016/j.shaw.2016.09.001.
12 Bottani, E., Monica, L., and Vignali, G. (2009). Safety management systems: Performance differences between adopters and non-adopters. *Safety Science, 47* (2), 155–162. https://doi.org/10.1016/j.ssci.2008.04.002.
13 International Organization for Standardization. (2022). ISO Survey 2022 results – Number of certificates and sites per country and the number of sector overall [Excel file]. Retrieved July 11, 2024, from https://www.iso.org/committee/54998.html.
14 See numerous articles written by Bruce Lyon and Georgi Popov, including: Lyon, B.K., and Popov, G. (2017). Communicating and managing risk: The key result of risk assessment. *Professional Safety, 62* (11), 35–44.

9 The Safety Market

Safety: Both a Profession and a Market

"We are drowning in information, while starving for wisdom. The world henceforth will be run by synthesizers, people able to put together the right information at the right time, think critically about it, and make important choices wisely."[1] If you have been recently assigned to lead an organization's safety program and/or lead in the development of an occupational safety and health (OSH) program, you will likely see, somewhat quickly, that there is both a large group of people engaged in the management of safety and an industry that surrounds them – as for many other professions.

Some of the groups and interests that one may see in this industry include but are certainly not limited to the following people and groups.

People supervising and managing safety or otherwise in OSH roles:

- People who take on OSH roles after completing an OSH-focused degree program
- People who become designated as responsible for OSH but from different backgrounds (e.g., human resources, environmental, quality, or production roles)
- People who are told they are to accomplish certain OSH tasks but are not given related job titles or "sanction"

Government and academia:

- People at universities and colleges who teach and conduct research in OSH
- Governmental organizations such as the Occupational Safety and Health Administration (OSHA) (regulators) and National Institute for Occupational Safety and Health (NIOSH) (researchers)

DOI: 10.4324/9781032704586-9

Organizations where work is done:

- Employers that have OSH issues, are subject to OSH regulations, and need related support and products
- People doing the work

Professional organizations and societies:

- The American Society of Safety Professionals (ASSP). This society provides development and representation of safety professionals and is engaged in a host of activities to promote the protection of people, property, and the environment – perhaps historically focused most strongly on occupational safety and health and on consensus standards development. There are many excellent standards made available through ASSP's leadership. It is a trusted resource for solid OSH information as well as an advocate for other important safety issues. See assp.org for more information.
- The National Safety Council (NSC). An organization dedicated to eliminating leading causes of preventable death and injury, historically incorporating a very broad view of safety to include the workplace, roadways, home, and communities. NSC has excellent resources for members, including library information. There is also an excellent consulting arm. It is active in providing education, including for people who may not have started their jobs in safety roles. See nsc.org for more information.
- The American Industrial Hygiene Association (AIHA). This is an organization that includes a focus on the health and safety of workers, as for ASSP and NSC above, However, there is also a focus on the technical aspects of the science and practice of industrial hygiene. Part of its mission includes community as well. It is also engaged in advocacy. See aiha.org.
- Other organizations that have a focus on OSH or things that contribute to a piece of worker safety. Each of these groups contributes in an important way. These include the American Association of Occupational Health Nurses (AAOHN), the Human Factors and Ergonomics Society (HFES), the American Conference of Governmental Industrial Hygienists (ACGIH), the National Association of Safety Professionals (NASP), the Association of Occupational Health Professionals in Healthcare, the National Fire Protection Association (NFPA), the Society of Fire Protection Engineers (SFPE), and others.

Manufacturers, suppliers, and distributors of equipment and products that support the industry:

- Personal protective equipment manufacturers, suppliers, distributors. This includes for things like safety helmets and hardhats, safety glasses, protective shoes, general protective clothing, respirators, hearing protection like ear

plugs and muffs, fall protection (full body harnesses and lanyards), life vests, flame-retardant clothing, etc.
- Other OSH-related product manufacturers and supply chains, including products related to machine safety as well as other facilities-oriented things, e.g., confined space entry equipment, spill control supplies, and labeling supplies.
- Designers, manufacturers, sellers, and distributors of safety-related measuring and monitoring equipment, e.g., multigas and other types of monitors for confined space entry, air sampling equipment, and supplies
- Emergency response equipment and supplies
- Software suppliers such as for management systems, EHS compliance, learning management systems, risk assessment and management, audit and inspection management, chemical and hazardous materials management, mobile apps, and others
- PPE detection systems and software using scanners and AI

Organizations that provide professional services:

- Providers of professional services, including both generalist and highly specialized consultants, laboratories, training associations, training organizations, and others
- Professional credentialing organizations
- Staffing and contractor vetting organizations
- Advertisers and marketers

Supporting industries:

- Workers' compensation and other insurance providers
- Mental health providers and resources

The overall safety-related market is not as large as some other types but is still significant and considered promising, with some exciting changes on the horizon.

Know that in the safety arena, there are many organizations and people having various areas of focus, coming from different backgrounds, and with different perspectives, goals, and objectives. Communication that you may receive and information you find as you research what you need to do to develop or enhance your safety program and fulfill your role for your organization could become overwhelming, confusing, and sometimes conflicting. If you are not careful, it could become difficult to see the forest for the trees. I mean, it could be difficult to see the real safety issues and things that *your* organization needs to focus on, compared to a potential avalanche of information from multiple sources.

The following section shares some ideas to keep in mind as you review multiple possible sources of information for your program.

Safety-related Societies and Organizations

As mentioned above, there are some national organizations dedicated to furthering OSH, with a broad view of the concept of safety, but with partially different areas of focus. They have a solid reputation and have existed for decades. This includes ASSP, NSC, and AIHA. There are also other organizations that can be relied upon for solid information, consensus standards, and advocacy, including some that focus on broader topics but that have an impact on worker safety, such as NFPA.

Organizations like these, consistent with their missions, provide a variety of helpful things. These can include, for example, professional development conferences and exhibitions of state-of-the-art equipment and products, the opportunity to connect and learn from others in the industry and in your areas of interest and specialty, training classes and courses of a managerial and technical nature to help you potentially do your job better, consensus standards, and literature helpful for understanding and addressing relevant OSH issues.

These organizations will often have a combination of training courses presented in different ways and at different locations. This can sometimes be through a core set of training topics and a stable of internal or contracted trainers, presented at conferences or through webinars and recordings. Training is sometimes also presented on other more variable topics given by either the organization's employees or by third parties that submit training and/or presentation proposals and that have been vetted. Both types of presentations can have great information and provide you with helpful resources and tools.

In addition to formal stand-alone learning opportunities, some organizations hold conferences at which volunteers from the organization's membership will give shorter presentations on various focused topics, carefully approved by the organization.

Also, some organizations provide periodic or ongoing free or for-fee continuing education courses that can be attended on a more flexible basis.

Some things to consider are that training topics will not always be what you need at a given point in time. Just because something is offered, emphasized, and maybe mentioned often in social media, it does not mean that the offering is right for you.

Of course, some organizations may present more courses in one area than another or essentially advocate particular approaches and types of solutions based on popularity or what is thought to be the latest and greatest information. Presenters can sometimes come from backgrounds that are based on one primary area of focus and injury prevention philosophy or another, or present certain types of information because that's how their full-time employer makes money.

If you would like to expand your knowledge in the OSH arena (because this could help you as you develop programs), you should absolutely take advantage of high-quality conferences, training, and networking opportunities. If your employer supports this, or it is a personal investment in yourself, ensure you pick from reputable options.

I would select carefully, based on a combination of what *you* need and what aligns with what you have assessed your organization to get the most value from. Conducting an organizational hazard and/or risk assessment can help with that.

Keep in mind that it can take years to become well rounded in OSH specialties or areas of focus. Some might not believe this when first asked to take on an OSH role, but "safety" is a broad field. If someone thinks it's simple and narrow – then they probably don't understand the broader profession. If someone intends to really excel in OSH knowledge and leadership, it is most effective to have a plan and improve knowledge intentionally and purposefully over time.

Safety Journals, News Magazines and Other Media

There are many different forms of media that talk about OSH either as a general topic or within an industry. Some sources of information can be very credible and helpful. Other sources can sometimes almost equate to just a collection of product advertisements.

Safety-related journals can be an excellent source of information. Content can include peer-reviewed articles on both deep and general topics.

Those and safety news magazines can share timely information regarding current issues and areas that are likely to be of interest to many. Reading or scanning them can be good to raise awareness regarding a variety of topics and give you ideas for things to include in your safety program's content.

Topically, you can often find content about specific issues that safety leaders may face, information about new and proposed regulations that can impact industry, highlights of top OSHA violation categories and penalties, new product information, and a host of other things. They usually also share information about upcoming conferences and training opportunities that you may want to review for possible attendance, as per the section above.

While magazines can publish excellent content, obviously editorial focus may or may not align with your needs at a given time. Fortunately, several magazines provide archived content, sometimes only to subscribers, that can be quite helpful.

Remember that authors can have different underlying agendas. Some may share information purely to be helpful to the broader safety community, i.e., they are intending to "give back" to the safety profession. Others, however, may be written with a slant that supports a popular theme or an underlying agenda. That underlying agenda may not adequately reflect other points of view, options or considerations, and may not be at all consistent with an approach you need.

Read those ads and other media content critically based on where *you* are in program development, and what *your* organization's context is.

Manufacturers, Safety Product Vendors, and Distributors

The occupational safety products and supplies market seems like a relatively strong and promising one. It can be good to keep up to date regarding things available on the market. It's good to know what solutions to safety needs and problems might exist. As you might expect, new and existing products are typically shown at the expositions tied to annual conferences and otherwise highlighted in safety news magazines.

You can learn a lot from these exhibits both in terms of getting ideas as well as becoming aware of direct solutions to problems your organization may have.

As a side note, however, I have rarely (but more than a few times) had someone try to tell me that a particular product was super impressive, that the OSHA inspectors in the region love it as a solution – it's OSHA-approved – and suggest that people would be fools not to buy what they are selling. However, OSHA doesn't approve products. Someone might have indicated that something seemed effective, but the OSHA inspectors I've known have always been careful about not endorsing products. It is obviously your organization's obligation to make sure that things meet OSHA requirements, and there are usually many ways to do that. Always leave yourself some room to investigate claims and statements about OSHA requirements that you may need to check.

I've also, unfortunately, had cases where a salesperson was not adequately versed in some aspects of a product they were selling, and described applications that did not seem safe. There are many well-educated, smart, and technically proficient salespeople, but as you are gaining familiarity with the safety concepts and requirements. be sure to ask all the questions you need to ask, and don't presume the sales representative always completely understands your application.

It can make sense to have good rapport with vendors who will respect your approach, be open to questions, and try to help you with solutions. Sometimes a vendor won't have what you need in the short term but is willing to stay in touch over time. I've had several excellent new products introduced to me that way.

Professional Safety Services

There are many types of OSH-related services that one can find. This includes safety consultants and training providers of various sizes, types, and levels of sophistication. Some provide services internationally, others nationally, regionally, and locally.

In an OSH context, consultants' services include things like interim staffing support, project safety management, on-site services (e.g., hazard communication program management), site auditing and inspections, gap analysis, risk analysis, special purpose projects (e.g., machine safety assessments, industrial hygiene monitoring, noise reduction), program development and assistance, cultural development and organizational change, OSHA compliance, due diligence for acquisitions, management and process improvement consultants, and others.

If your organization has a need for safety improvement in some way, there are many options to consider for engaging consultants. As for other categories of safety services – if you are evaluating options, start by having an idea of *where* your organization is at and where its needs are. Although, certainly, consulting organizations can help to identify compliance and best practice gaps, provide system recommendations, assist with program development and rollout, and help with much more narrow and specific services that you may identify and define.

When it comes to larger management system projects and projects related to safety culture and injury reduction and minimization, there are a few different schools of thought. There is not just one standard way that a consultant might approach helping you, nor one mental framework for what works best for injury prevention.

If and when you select a consultant for services, pick critically. Some firms are well known for focusing on certain practice areas or industries and can have varying reputations. From an OSH perspective, consider factors such as the company's relevant experience with whatever it is you need assistance with, its capacity to take on new clients and projects of certain sizes, digital skill sets as it may be important, and fit. As an example regarding fit, you could be in fundamental disagreement with the approach that a consulting organization advocates for you.

Regarding training consultants and providers, there are usually several options available to you, depending upon the topic and need. Training is available both to the individual safety leader and to organizations that need training for special initiatives or standard OSHA-required topics. Larger safety associations offer some forms of training. There are also universities that offer some continuing education–style courses, OSHA outreach training centers, other industry-specific organizations that provide training, and independent providers. The quality and utility can be highly variable; ask questions like usual!

Course options are plentiful from a topical standpoint; the trick can sometimes be finding a course at a time that fits your schedule, if you are looking for in-person/live training. That can be especially true for weather-dependent training.

Course titles can be deceiving. Some make it sound like attendees will become masters of everything and also receive prestigious-sounding course certificates. Put some extra effort into ensuring that course content will include what the title suggests.

Safety-related Software and Web-based Tools

Safety-related software programs are a big deal. There are a variety of programs that can address several needs and help organizations operate more efficiently. Good software can be essential in certain applications. I've reviewed some excellent systems. Users may be quite happy to have a system that can manage a broad range of functions. Some systems are designed for an enterprise, and some for smaller-scale applications with a limited range of functionality. Costs can be highly variable.

Don't be dismayed to know that to make some systems work for broader applications and for companies with unique processes, significant time and energy may be needed to tailor and roll out programs. That can include to develop documentation, provide training, and debugging.

There are some fabulous products capable of doing a lot, but organizations need to be careful in what they pick. They should also expect to put in dedicated time and effort to make things work best.

On the less comprehensive side of things, there are providers of inspection check sheets for use with portable devices, which can be modified to fit your organization's projects or processes. Those can be very helpful, easy to use, and assist with communication. Sometimes extra training may be required for employees to understand what some of the questions are asking.

Of course, inspection and audit software in and of itself helps with just one part of what many organizations need to have in their safety program – unless that information is designed to integrate with other program elements.

What You Can Do

There are a couple of things that you might want to think about as you process information from the multitude of sources you will likely receive it from, and apply it to your program development effort.

The first thing is that if you conduct a risk analysis/risk assessment, by yourself, with a work team, or with the help of a third party, you will have an idea of where your gaps, risks, and opportunities are. That information can provide you with some framework and something to work from. It can help you determine where training needs exist, provide information that you can compare to and align with business objectives, and know what things might be more important than others on a risk-prioritized basis.

The second thing is that you can and should apply critical evaluation of what you are presented with in the safety arena. There are many competing forces and agendas at play. Even well-intentioned people can attempt to steer us in directions that don't best match our needs.

The risk assessment mentioned above can help you to know what your organization needs to work on, but not the best methods to address the identified risks and opportunities. Many options and potential solutions may be available.

When evaluating marketing information and options for creating or improving your OSH program, consider the following:

- Is what you are thinking about doing consistent with the organization's hazard identification and risk assessment processes? Will it apply resources where they are needed based on values and priorities?
- If the improvement is implemented as planned, how will that change the organization's compliance and risk profile?
- How might it impact or align with other processes?
- Is the solution validated and well trusted or something that is not yet proven in your context?

- Will the methodology or changed processes fit your organization's culture, readiness, and maturity?
- Will it help shift the organization's approach from bare minimum "compliance-only" to a stronger approach?
- In what ways might this support and align with other values?
- Will employees have an opportunity to provide input to process/change ideas?
- Are the changes consistent with human factors design concepts and will it make processes easier for workers?
- Could implementation of this detract from other higher priorities or wind up being a "flash in the pan"?
- Could the product or service cause problems elsewhere?
- Will physical changes impact other regulatory requirements, and if so, how can that be managed (e.g., potential impacts to waste streams and air permitting requirements)?
- Will additional training be needed for implementation, either initially or in the long term? Can this be provided?
- Will this really help your overall safety processes, or might it deceptively support "perceptions" of safety yet provide only a patchwork?
- For consulting, does the organization have a proven track record of success for similar relevant projects?
- Will planned new processes result in some longer-term obligations or reliance on a third-party provider of something? If so, how can that be managed and changed over time?
- Will changes help promote worker involvement and engagement and an otherwise positive safety culture?
- Will you be able to describe the financial and the intangible value that this will bring to the organization?
- How can the timing of this improvement align with the organization's anticipated cash flow?

There is a large OSH community with many dedicated to improving worker safety. However, the range of potential safety issues and needs is broad and, of course, no one person knows everything. Your organization has some things that it will need to do and that only it can do for itself. However, there are many options for how organizations approach their OSH programs. Reach out to learn what you need to learn. Be open-minded but evaluate the information you receive carefully. It is suggested that you use some form of risk-based approach to sort through what is most important and create a game plan for your approach to OSH management.

Note

1 Wilson, E.O. (1998). *Consilience: The Unity of Knowledge* (p. 294). Alfred A Knopf.

10 What Have Others Done?

Some Learning Examples

From *Theory* to *Doing*

"Greatness is primarily not a function of circumstance. Greatness, it turns out, is primarily a matter of conscious choice and discipline."[1] Earlier chapters discussed a lot of the background behind why organizations may or do need safety programs and plans, and the benefits of having them. The prior information was about the "whys" and where you can find information you will need, based on your organization's specific situation and the safety market in general. We reviewed a few things that you'll need to think about as you craft your program, such as document structures, staff responsibilities, and program rollout methods. Information about occupational safety and health (OSH) programs versus OSH systems was included to present the option of using an intentional *systems thinking* and *risk-based* approach in what your organization rolls out. Occupational health and safety management systems (OHSMSs) are not explicitly required by the Occupational Health and Safety Administration (OSHA), but the use of one can be helpful in many (but not all) cases. In the end, the intentional and careful choice of safety program structure and elements and a disciplined approach in application can be important factors in OSH program success.

With that information in mind, we can talk about the "art" of assessing what an organization needs either for a new program or for an enhanced or revised one. In this chapter, we review how two different organizations worked to meet OSHA requirements through a process of program development in one case and program improvement in another. The cases illustrate different levels of conscious choice and discipline; program results differed as well.

Discussing these two examples helps illustrate and reinforce safety program/plan concepts.

Approaches to developing a program can vary depending on need, and there is no one-size-fits-all approach. We discussed a number of these factors previously, which include internal and external factors that will or can affect the organization's OSH environment, industry, timing, organizational size and maturity, nature and location of work, culture, and many other things.

There are a few implementation thoughts that I believe are worth emphasizing here. While organizations are different, and the attitudes and skill set of

DOI: 10.4324/9781032704586-10

individuals are different, consider the following when assessing operations that require safety programs:

1. If the organization shows it truly "cares" about workplace safety and is not just dotting "i's" and crossing "t's," that can mean a lot to the workforce. That can do more for your program sometimes than what written program documents say on paper.
2. Try to make the program a bit fun and different, where you can. I've seen that make a difference for engagement, buy-in, and interest.
3. For some types of programs, literal and consistent compliance with procedural safety steps is crucial. However, we need to concurrently remember that taking an overly legalistic approach to safety can result in missing important signals about our programs and things that need to change. Write programs with end-users in mind.

In the following sections, I describe some of the process that two different organizations went through when creating or updating programs. I will anonymize these organizations by referring to them as Organization A and Organization B and changing some otherwise identifying information. What these companies did will obviously not be relevant to all organizations. However, you might be able to grab a few ideas from the summary.

These examples focus on safety and health considerations, consistent with the theme of this book. It is recognized that many organizations have staff that address both OSH and environmental requirements, depending on a variety of factors. Those environmental requirements can sometimes be combined with an OSH review process, or vice versa, depending upon areas of expertise and how the organization divides up responsibilities. Sometimes it is appropriate and at other times not. Ultimately, the people performing assessments should have a skill set commensurate with the type and level of hazards that may exist or could be created.

Organization A

This organization was engaged in starting up a new manufacturing facility at a location in the United States. It was a part of a larger organization and could take advantage of some resources available through the parent company but was responsible for creating a mostly independent facility. A core group of employees of the parent organization was assigned to create and lead a small team of others (including new hires) through a full design, facility construction, equipment and process installation, hiring, and production kickoff.

For some processes such as for advanced manufacturing engineering, there were more resources available to the kickoff team than for other new plants. However, the team also utilized many OSH approaches that could be used by most organizations.

Processes for Program Development

Organization A had a parent corporation that specified much of the technology to be used in processes to be installed at the new plant. Even so, the small core team and newly hired staff were responsible for site considerations, including safety. Different staff focused on different "functional areas," but the team overall had responsibility for both traditional plant safety issues as well as those safety issues tied to new technology and equipment.

More traditional safety items pertained to the review of planned facility layouts, grounds, powered industrial vehicle traffic, receipt of raw materials, transportation of materials to various manufacturing lines (as input), and transportation of finished products to a warehouse and ultimately shipping processes. Beyond that, other planning was needed to understand new equipment from multiple manufacturers and their related hazards, automated motion between work cells, numerous robotics cells and conveyors, as well as cranes and other material handling devices. There were also specialty work areas within the plant, such as a laboratory, maintenance, welding, and a tool room. All these needed to be addressed as part of safety program development. Corporate-level procedures were available for possible use but not mandated, except for a multi-step new equipment review process. To a large extent, the parent company advocated a "fresh" approach for the new plant's processes.

What Was Reviewed

As you might imagine, the program development effort for this organization was just that, an effort. It did not happen overnight. Fortunately, a team was engaged early enough in planning to be able to have relevant processes in place before early physical work, and programs were able to be developed that were flexible as the organization's processes and numbers of employees grew.

Consideration for safety started in the conceptual phases of design. That included for physical construction of the facility as well as for processes and equipment. The inclusion of safety continued over time to include aspects of facility construction (from the manufacturing workers' perspective), as well as during and through new process and equipment testing and installation.

This was intended to help determine what needed to be included in the organization's written OSH program in comparison to OSHA's written program requirements, as well as to help ensure that things being built and installed would be built and installed in a manner that met standards and codes (not just OSHA requirements), and which would be safe to operate.

"Safe to operate" included consideration for placement, integration with other equipment, worker interaction with the machine and processes, and safety for visitors and suppliers. It also included gathering information that would be used in the development of equipment energy control (lockout/tagout) processes based on operational needs, compliance with OSHA requirements, and the identification of initial and longer-term training needs.

The following bulleted list captures some of the things that were reviewed or actions that were taken as part of an initial and ongoing assessment process. This list is not comprehensive. The items listed are not in strict chronological order; reviews were cyclical over time based on staff availability and installation timing.

- Review of facility drawings/plans
- Review of project management charts showing planned construction and installation timing
- Review of documents that summarized new plant scope and planned manufacturing, assembly, and subassembly operations
- Consideration of approximate hiring plans and timelines
- Review of workforce OSH obligations pertaining to direct-hire versus temporary employees
- Consideration of communication, safety, and potential emergency obligations with third-party employers
- Receipt of construction and process installation status updates via meetings and site walks
- Review of similar (but not usually the same) processes at other company facilities
- Discussions with engineers/planners: how they envision work, perceived risks, and known problem areas
- Review to ensure appropriate facility lighting, suitability/audibility of alarms, process and pipe labeling, hazard warnings, potential exit routes for planning, potential emergency sheltering areas
- Implementation of "soup to nuts" new equipment review, installation, and buy-off processes (consistent with elements of what is referred to as "prevention through design")
    - Involvement of persons with safety competence in design phases (e.g., 30%, 60%, 90%)
    - Creation of equipment review checklists based on type of processes/machines that served as a memory jogger for potential hazards and program needs, and for documentation of status and progress
    - Visits to key equipment suppliers prior to equipment delivery
    - Interaction with engineering team during planning, installation, and commissioning processes (regarding questions, concerns, and items needing correction)
    - Consideration of interface between workers and machine processes, including for ergonomics, access to areas of the machine during production, maintenance requirements (which differ from normal operation), and other access needs
    - Review of things inherent to the machinery and configuration between machines, such as space between machines, conveyor and robotics interfaces, openings that could present pinch points, guarding needed to prevent access to hazardous motion, internal locations likely to

require frequent access, and/or locations with hazards needing safety controls

- Inquiry into various operating configurations and different operating conditions, using *what-if* questions: "If the process is in automatic mode and if …" "If the user puts the machine into manual mode pushes on …" "What are the pressures and temperatures that the process …"
- Review of features required by the organization's specifications, such as for safety relays and control reliability (this can be very important to some approaches to safety)
- Review of the potential for the unintended release of hazardous energy for inclusion in energy control programs, multi-worker access safety, electrical considerations (main electrical panel interlocks, interior finger-safe protection, arc flash prevention and labeling), emergency stops
- Review of other potential issues such as hazardous noise, potential creation of exposures to hazardous substances, fire potential from product build-up, need for access from heights (potential fall protection needs)
- Review of processes for potential personal protective equipment (PPE) needs, which would feed into the site's overall PPE assessment/certification
- Review of suppliers' equipment risk assessments where available
- Approval of new equipment in stages, before full release to production

- Review of chemicals planned for use and materials that may otherwise be consumed. This could impact control needs, hazard communication program needs, the need to ensure a safe atmosphere for workers, and the need to ensure that the materials would not create combustible or flammable atmospheres.
- Review of hazardous materials planned for use, which is tied to processes for material approvals, receipt, handling, storage, and use
- Review of chemical storage areas for things like spill containment, compatibility, ability to detect and control leaks, and the potential creation of chemical exposures
- Review of (lifting-related) material handling needs, locations, capacities, inspections, and corresponding training needs
- Review of facility and possibly process-specific fire suppression needs
- Review to ensure emergency shower/eyewashes as needed
- Review of intended hazardous process areas such as designated welding areas and areas of rapid activity for assurance of appropriate processes and approvals
- Review of walkways with respect to powered industrial vehicle traffic, visibility, head-bump potential, tripping hazards, sharp edges, means of access, and egress
- Review of work platforms for fall prevention considerations

- Tool room reviews (for common shop hazards, access control, and training needs)
- Laboratory reviews (especially regarding common lab acids and adequacy of fume hoods)
- Review of powered industrial vehicle needs, types, battery, and/or liquified petroleum gas (LPG) considerations, including PPE and training needs
- Review of mandatory OSHA-related postings and notifications
- Review of various in-plant locations for possible permit-required confined space entry program applicability
- Comparison of regulatory requirements, including OSHA-required written programs/plans versus anticipated processes

How Did Teams Work toward Development?

The site team worked in an integrated manner and with the understanding that regardless of what functional expertise one had, all members were expected to support safe work and the creation of safe processes.

New equipment review processes were conducted in an iterative way, by people with expertise in safety, engineering, operations, and maintenance functions. This was of significant value and ensured that multiple perspectives were considered in reviewing equipment before release to production.

Progress review meetings were held routinely and provided an opportunity to communicate about problematic areas, potential process delays, and sometimes modified areas of focus.

As was intended, the early review of plans and activities contributed to the identification of needed written programs and also provided the opportunity to install equipment that was designed and configured to operate safely.

Appropriate procedures were developed in an appropriate order. Some documents started off being written somewhat generically and were refined later as installation and hiring progressed.

The equipment buy-off process was of tremendous value. Operating under a common goal, it provided a means to obtain multiple perspectives using a problem-solving approach. This included the ability to obtain input from early hires regarding their perspectives and concerns, where they existed. It also created a process to request safety-related machine modifications, where necessary, before doing so became exceedingly expensive.

The creation of new processes and procedures was considered an opportunity to positively influence safety culture as new staff were brought on board.

Issues Encountered

While most equipment suppliers attempted to meet the organization's equipment specifications in good faith, there were still some that presumed the facility would not compare equipment specification requirements to what was actually delivered. While some specifications might have been a bit more esoteric, others pertained to things like emergency stops, noise levels, and the release of airborne materials.

For Organization A to have provided specifications was, overall, of tremendous worth and important for consistency. However, sometimes equipment as inspected during installation did not match the stated requirements, requiring in-plant modifications. In yet other cases, equipment suppliers had presumed that an approval in the European Union would suffice, whereas the approval did not translate to what the specifications actually required.

The point here is that equipment manufacturers may not always provide your organization with requested functions or foolproof configurations. There is no such thing as an OSHA-approval process for manufacturing equipment to which a supplier must comply before selling the product, unlike in some other countries. This is discussed in more detail in Chapter 11.

On the contrary, your organization has the obligations we have discussed previously under the OSH Act, Section 5(a)(1), and an obligation to ensure that processes meet other specific OSHA standards that may exist. Also, if your organization is subject to the OSHA Process Safety Management standard located at 29 CFR 1910.119, then it has certain obligations to conduct process hazard analyses (see the standard).

Results

The processes described above were used to identify the need for and create OSHA-required written programs, as well as determine other program needs. They also helped to identify issues needing correction related to new equipment installation, which were addressed as part of Organization A's new equipment review and approval processes.

The activities also helped the site to create processes not directly mandated by OSHA but that helped the organization to meet OSHA requirements. This included procedures such as for operational change control, health and wellness–related activities, ergonomics considerations, well-defined safety committees, and management review of critical safety processes.

The site had created well-functioning and successful safety processes through a combination of written and unwritten programs, as well as its manner of program implementation. Over time it shifted from activities associated with a new facility and new processes to other sustaining processes, and even a few things that helped make safety processes a bit more fun.

Organization B

Organization B was a small manufacturing employer that was expanding its site footprint. It had one small production building and warehouse. Processes had included the receipt of raw plastics through silos that were fed into molding processes and extruders. There was significant manual interaction with machines to set up, make, eject, and manually trim products, along with other associated processes. Finished products were stored in the warehouse and shipped to customers via a third-party transportation service.

The organization had requested a safety gap analysis for the purpose of ensuring its safety programs were OSHA-compliant, in view of its significant planned expansion (and local press coverage).

Processes for Program Development

Organization B was classified as a small business. It had some elements of a safety program in place. It did not have a person acting in a safety leadership role. The review was conducted by one consultant.

What Was Reviewed?

Manufacturing processes were already in full operation. There was no documentation available to review equipment that would be added as part of the expansion planned for later. It had been stated that new equipment would be similar to some of the existing equipment.

Assessment of this organization was conducted via site walk-through, interviews with workers, and a review of available documentation versus OSHA's requirements for written programs.

Items reviewed and actions taken had included:

- Review of injury and first aid records, insurance company reports, and prior OSHA citations
- Review of safety policy/"handbook" and training materials
- Review of older, draft programs
- Walks and discussions through production area, warehouse, shipping/receiving, raw product receipt and storage, external facility areas
- Observation of work in all areas
- Discussions with employees regarding responsibilities, what they do, and potential items of concern
- Contextual review of programs related to accessibility to silos and corresponding fall prevention/protection, confined spaces, energy control (lockout/tagout/tryout), equipment training, emergency action and fire prevention, first aid programs and supplies, flammable materials storage, hand tools, hazard communication, hazardous noise, machine guarding, PPE, powered industrial vehicles (fork trucks), and workplace violence prevention.
- Review of observations and available existing programs compared to OSHA's requirements for written programs

How Did Teams Work toward Development?

Site visits showed limited supervisory staff and that managers and supervisors were assigned many responsibilities already. This is not uncommon to smaller organizations. No one felt comfortable taking on additional responsibilities, nor did they have significant prior OSH training or experience. No one had been assigned to take a safety leadership role (e.g., no safety coordinator or manager).

The company had provided excellent operator training for some processes. However, not surprisingly, written program and "other" gaps existed that needed correction. Based upon the items reviewed when compared to OSHA's written program requirements (see Chapter 5 and OSHA's online tools), written program needs, as well as other necessary actions, were clear.

The organization was provided with a report that summarized written program requirements (gaps), other hazards requiring action, best practice recommendations for program structure and management, and information about further resources.

Recommended actions included suggestions for overall safety program improvement, including enhancement of roles and responsibilities. There was no OSHA regulation that specifically spoke about "ownership and accountability" in this context. However, without addressing the concept, it is not always easy for organizations and workers to know, clearly, who is supposed to do what or to make sure that required actions take place.

The report also recommended processes for hazard identification and the tracking of open issues. There was again no explicit OSHA requirement for this. However, as discussed previously, employers have obligations under the OSH Act, and the location had a larger number of issues to work through.

As a result of the report and recommendations, the organization decided to demonstrate a renewed management commitment to OSH and program improvement. It assigned an individual to take a lead role for safety programs management, while also providing additional leadership level safety training to a few people who were to assist with implementation.

The organization also re-established a safety committee but with the addition of a clear scope and goals, through which it managed its improvement processes. Missing and inadequate written programs/plans were developed through this process.

Issues Encountered

A primary issue had been that no one in supervision felt comfortable taking on additional duties, nor felt competent in the understanding and administration of OSHA requirements. However, the organization was not large or complex, and safety-related issues and requirements were relatively straightforward. Workers became more comfortable in their roles through extra assurances and support from executive management, limited temporary external assistance, and the receipt of selected external training regarding OSHA's requirements.

Results

The OSH management changes described above were adequate for the organization to comply with written program requirements, correct other identified gaps, and have a process for the evaluation and control of future safety issues that might arise. The organization felt that it had successfully reduced its risk.

Summary

The scenarios described above were provided to give real-life examples of working through an OSH program development process.

Of course, these examples cannot capture all the considerations that you might need to think through as you develop your written and unwritten programs. There is good information available for those needing to develop programs, including easy-to-access templates. However, each organization still needs to assess what processes and hazards they have for which written safety programs and processes are required, and what else they might need or want to supplement their mandatory written programs.

That all takes some work, if an organization wants to create and implement meaningful programs that will be used, and obtain related benefits. Making a conscious choice to obtain those programs and applying discipline in their application can result in great success.

Note

1 Collins, J. (2017, January 1). Greatness is a matter of conscious choice and discipline. *Jim Collins*. https://www.jimcollins.com/media_topics/GreatnessIsAMatterOfConscious.html.

11 Consider Prevention through Design (PtD)

The Concept of Prevention through Design (PtD)

An ounce of prevention is worth a pound of cure.

Benjamin Franklin[1]

I am writing this chapter with the intent to save you some grief, and help your organization to reduce risk, avoid unnecessary injuries, and minimize long-term costs – through something that can sometimes be achieved with only nominal or moderate effort.

What am I talking about here, specifically?

I am talking about two things:

1. Ensuring that new and redesigned processes, equipment and facilities are designed and installed with safety in mind, before something bad happens. I am referring to both machinery and process safety as they relate to worker safety.
2. How construction projects can be planned in order to enhance worker safety, and designed in a manner that promotes longer-term safety for those that use and maintain a facility.

Manufacturing-oriented processes, equipment, and facilities

I've seen many organizations build new facilities, install new processes and machines, and make changes that can impact worker safety – without giving much or any thought to "safety" before buying equipment or embarking on work. And then later discovering serious issues related to how something was built and/or installed that increased the risk of injuries from use of the new process.

Unfortunately, when things are discovered this way, after the fact, it is often the case that work is too far along to "fix" anything feasibly without major redesign and rework, and project funds are almost gone. (How many people in safety roles have heard, "So sorry, I thought that was good enough"? Or, "Maybe we can fix

DOI: 10.4324/9781032704586-11

it after we make some money off this line"?) While acknowledging that projects have budgets and that project managers have an appropriate desire to minimize costs and maximize returns, organizations should have processes to ensure timely and adequate consideration of regulatory requirements, injury potential, and other safety-driven impacts on the organization.

It is best and most effective to identify occupational safety and health (OSH) related needs (and environmental ones) early in the conceptual stages of process design and then later in progressive design phases. Otherwise, once there is significant planning time, research time, and development of a mindset about how things are going to work, including how process pieces will work together (how does "this" part fit with "that" part, how does part "x" feed into machine "y"), it can be much more expensive to make changes. Retrofitting, where it can (practically speaking) be accomplished can be an expensive way to go.

Different organizations, of course, have different process and machine needs. In manufacturing, at least, some organizations can have complex systems, and others more simple ones.

Examples of the types of problems that can occur without adequate (and early) design consideration may include the following:

- For most organizations: A mismatch between the human worker, their physical dimensions and strengths, and the way they think, act, assess risk, and expect processes to work, with the machine or process as delivered and installed
- For most organizations: Lack of timely knowledge of inherent hazards of operation and operating instructions, to ensure proper training
- Risky physical demands that go beyond what can be expected without injury; designs and cycle times that are not compatible with the best ergonomic principles of design; lack of adjustability for the worker
- Lack of adequate material handling equipment to prevent ergonomic injury, cranes and hoists that are not selected, inspected, and installed appropriately to prevent crane failure or dropped loads
- Unsafe walking/working surfaces, inadequately constructed ladders, work platforms, catwalks (especially regarding lack of fall prevention)
- Cut hazards from sharp edges on machined parts or the intersection of guarding structures left rough (e.g., hollow structural sections [tubular frames])
- Hazardous noise emissions due to inadequate noise reduction controls
- Lack of, or inadequate, guarding to protect from hazardous motion, pinch points, etc.
- Designs that make routine and periodic work tasks difficult to perform
- Difficult-to-access (or lack of) locations for the de-energization and control of hazardous energy when conducting work covered under the Occupational Safety and Health Administration's (OSHA) "Control of Hazardous Energy" standard (29 CFR 1910.147).

- Chemical storage violations (related to quantities, location, type of storage); inappropriate dispensing (static discharge risk/bonding and grounding issues, proximity to energy sources that could cause ignition)
- The creation of unsafe atmospheres due to the emission of toxic and/or combustible or flammable materials.
- The lack of devices to remove exhausts, fumes, etc., where warranted
- Failure to meet a variety of electrical standards
- Undetected/unknown potential failure modes
- Inadequately selected engineering controls and other risk reduction methods, inadequate functional safety devices and protective measures; lack of controls that meet desired levels of protection
- Placement next to pre-existing processes in an incompatible and unsafe way
- Other life safety code violations (e.g., old sprinklers are no longer appropriate for the hazard, inappropriate extension cords, blocked paths of exit)
- Machines with significant environmental risks

If you have significant machinery safety considerations, a great reference is the consensus standard ANSI B11.0,[2] which talks about machinery hazards, risk assessment, risk reduction measures, and many other things. Annex D includes a section titled, "Partial List of Machinery Hazards" that can be helpful.

The ANSI/ASSP Z590.3 consensus standard on PtD[3] describes a comprehensive approach to PtD. If your organization designs (or has designed) and brings in new processes or redesigns processes or systems that can have safety implications, then the Z590.3 standard can be an excellent reference for your safety program or system. It uses a complete life cycle approach. The standard includes but is not limited to discussion of design safety reviews, management of change, risk management processes, and even techniques that can be used in various parts of a broad approach to risk identification and assessment, along with many supporting and informative addendums. It includes references to "construction" of facilities, but primarily with respect to processes and systems within.

If the organizations that I mentioned above had considered "prevention" in their design work, or communicated with its contracted sources regarding the avoidance of safety issues when planning and the design work was starting, problems might have been avoided.

PtD processes *can* be in place for new designs so that intentional checks are conducted to be sure to avoid OSH problems; to evaluate what could go wrong with the new or redesigned system in question – using safety-focused review processes; or to check if installation is going according to plan and to see if there are other hazards not previously identified that can be addressed during process integration.

A little effort up front can go a long way. However, many organizations don't have significant in-house resources, and even some that do presume that if they are paying someone else a lot of money to supply equipment or tools or build something, they are safe to assume that what they are buying is designed and constructed properly for safe operation.

That assumption is not always a good one!

I am not an attorney and won't go down the road of discussing what builders, suppliers, vendors, and manufacturers are or are not responsible for from a legal perspective.

However, from the standpoint of what OSHA does or does not do – planners, purchasers, and safety staff need to know that there is no such thing as an "OSHA approval" for industrial machinery or equipment. You may see some product literature or markings on products that indicate something is "OSHA-approved," but OSHA does not test, certify, or approve things. There are requirements in some regulations that certain products be tested by Nationally Recognized Testing Laboratories (NRTLs),[4] but OSHA does not perform that function. Sometimes sellers may be attempting to imply that the way something is built or constructed is compliant with what OSHA would require, but again, there is no OSHA certification process.

Unfortunately for end-users, I've seen many cases where a supplier will low-ball their equipment sales proposals and tell customers that if they want more guarding or labeling, or anything else, that they can accommodate that but will add fees. Yet their starting point is sometimes deficient or very questionable at best.

And from an equipment supplier's perspective, while your organization's planners may know conceptually (at least) what machines and equipment are going to need to do, suppliers may not automatically know your organizational needs, installation specifics, or nuances of use. The machinery safety *consensus* standards available through ANSI in the ANSI B11 series describe risk assessment processes and the obligations of manufacturers, users, and others. But again, these are not OSHA standards and are not incorporated by reference into the OSHA standards.[5]

Construction Project Design and Planning

Another area of focus in the PtD world is in the construction and constructed facility realm. The idea is to maximize the safety of construction workers during construction processes, and to consider the safety of facility end-users such as for maintenance processes. For example, the order, ease, and manner of construction can impact workers' safety. Builders can also select certain means and methods that can be safer than others. The concept in general has been an area of emphasis in construction programs, in the Center for Construction Research and Training (CPWR) (historically, NIOSH's research arm for construction safety), and elsewhere. Information about PtD has also been shared by the American Society of Civil Engineers (ASCE), including considerations relative to US-based construction and project delivery methods.

The concept applied to facility end-users encourages designs that can make long-term maintenance and use of a facility safer. That can include prevention of falls by designing access a certain way, providing permanent fall protection anchors, and other considerations.

If you work in an architectural or construction-oriented firm, then application of PtD concepts may have a different focus for you than for, say, manufacturing organizations with an industrial equipment or processes focus. However, organizations with different missions can work together to apply PtD concepts collaboratively. If you are working for an organization that is or will be building new facilities, you can think about PtD applying not only to your equipment-related processes but also to the longer-term safe use of your facility. Don't presume someone else is going to do it for you or that they will think through things in the same way that you might as the end-user.

What Does This Have to Do with Writing a Safety Program?

PtD has been a concept intended to improve processes that can impact safety but has not been a regulatory requirement in the United States. However, the application of PtD concepts supports obligations for employers to provide a safe workplace.

Incorporating PtD principles into the way your organization approaches safety can help it to not only comply with regulatory requirements, but also to significantly improve the effectiveness of its efforts. Also, if your organization decides to implement an OHSMS, using PtD processes will support key OHSMS requirements like hazard identification, risk assessment, and management of change.

Exactly what and how you might apply PtD principles can depend of course on the type of organization you are in and what its mission is. Are you responsible for multiple sites or one? How often does the organization design or purchase new facilities, processes, and equipment? Do you procure custom equipment or largely purchase things "off the shelf"? What are your organization's internal capabilities? How much leverage might your organization have with its suppliers?

To the extent that your organization can benefit from having a process to ensure that possible risks are addressed, applying even a few PtD processes to improve *something* is probably better than *nothing*. Sometimes a key question might be, "Is your organization willing to have the conversation?"

My suggestion is that your organization include at least basic PtD concepts in your safety programs commensurate with your OSHA obligations, feasibility, and risk tolerance (or, if you implement an OHSMS, into your safety systems). As discussed earlier, organizations have obligations under OSHA to provide a safe workplace and employee training. However, there are a range of options for how an organization can "get there," and different organizations have different inherent process risks and internal capabilities.

An organization's PtD-aligned processes could include things like:

- Having a change management process. I am referring to having an internal process that requires that organizational changes that could impact safety be reviewed to determine if planned changes will be safe and if the changes themselves could cause other hazards. This could pertain to things like changes in machine integration, guarding, and safety controls.
- Communicating certain design or performance requirements from the organization to the appropriate vendor (for equipment, for facility modifications, or other things). I've seen these come from procurement (if the organization has that function), engineering, or other parts of organizations. While you may not have access to high-level "manufacturing engineering" resources or a strong understanding of applicable technical standards as larger organizations often might, it might still be possible to convey safety-related requirements.
- Developing and using a checklist or other risk assessment techniques to help identify hazards determine whether the hazards pose risks, and determine whether *action* is needed to address the risks. The Z590.3 standard reviews some options. A review of the ANSI B11.0 standard covers a lot of topics that can give you ideas. It's getting easier and easier to find general resources on the internet, too. Your organization could consider developing some relatively simple checklists that call out safety requirements, along with having a process to think through what types of things could go wrong. If you can build the organizational expectation that it will consult with subject matter experts if needed, and if you can identify and address some *human being*–type considerations that can cause errors and/or injuries, why not do that?
- Having a machine commissioning or buy-off process that requires that new processes be reviewed for compliance and other risk factors, before approval. Some organizations have review processes that require a team of people with different knowledge and skill sets to be part of that process.
- Having processes that require a *safety buy-off* for new facilities and modifications to facilities. This can facilitate taking potential risks to a facility's end-users into account, which could pertain to space in working locations, working at heights, methods to access systems, and other things.

Common Misconceptions in Purchase of Equipment and Processes

I've met a number of people over time who don't understand the value in having processes consistent with PtD. Some have asked, rhetorically, if it was not the case that they can count on manufacturers selling industrial equipment to meet OSHA requirements.

It can be easy for some to suppose that. However, as before, OSHA does not play that role. Other countries do have enhanced liabilities for companies selling equipment and perhaps in international organizations, which leads to some confusion.

How much can one rely on their equipment supplier to provide "safe" processes? I think that probably depends on production volume, complexity, and uniqueness of what is being sold. Sometimes an organization probably has a strong basis for relying entirely on a manufacturer and does not have internal resources to investigate otherwise.

However, you might be surprised at how many times I've visited suppliers as part of PtD-styled equipment buy-off processes and found significant issues that either absolutely need correction or were otherwise not optimal.

There are all types of people and companies that make things. They are not all equally aware of OSHA or consensus standards, and they don't all have the same ideas about what poses a risk and what doesn't. As an aside, have you ever seen things advertised online through short videos as if a product or work method were miraculous and hugely time-saving, and you've thought to yourself, "that can't possibly be safe," or "you've got to be kidding me?" Well, what is the reputation of your supplier, and how badly do they want to sell you their miraculous new equipment?

What Can Success Look Like?

I do not want to leave you with the impression that every organization needs to have a super elaborate process. Not every organization fits that mold. It really does depend on many things.

For some organizations, it can be an accomplishment to simply ask some basic questions and check to be sure that a few things are considered. They may have very little leverage with their vendors but can at least be aware of potential issues and have a simple process to make the best purchasing and in-facility decisions. For other organizations, it could mean having a global and sophisticated team with many millions of dollars in capital budgets and large, diverse teams.

Fundamentally, is your organization willing to take some reasonable steps?

The Business Case for Safety

With all that said, it can still be the case that equipment or processes are supplied and installed but hazards are discovered later. Sometimes an organization will discover that something is not compliant with OSHA requirements and it will take significant work to correct. At other times, it might discover that configurations don't meet consensus standard requirements that the organization had committed to follow.

Or it might detect something not directly regulated but that it would like to do something about. That could pertain to human interaction with the process that requires unsafe lifting, bending or twisting, excessive forces to various parts of the body, a lack of local exhaust ventilation from processes emitting vapors or fumes, the surfaces that someone works on, walking and working surfaces (platforms, ladder, etc.), or other significant things. Sometimes it might be hard to identify hazards unless you view the actual performance of job tasks.

If risks were missed, organizations still have a fundamental obligation under OSHA to provide a safe workplace and to comply with OSHA standards.

At that point, people in OSH roles may need to act, but work in an environment with no process that defines what to do or how to communicate issues and needs to the organization.

Sometimes the missed items can be called out and addressed as part of an existing budget and be corrected with not much difficulty.

However, funds can often be expended by the time an issue is discovered. In those cases, you might be asked to prepare a cost justification, some sort of financial justification, or make a capital budget request that provides a description of the basis and cost justifications.

My viewpoint on this is:

1. You are now in an unenviable position. It usually takes more effort to justify a retrofit than to have avoided the problem to begin with. The organization will likely have higher costs to deal with, along with possible production delays and decreases in production volume. The ease and method of making changes … changes. Usually, more paperwork and a different set of approvals are needed to correct the issues.

   It can often be the case that while the organizational decision-makers and/or purchasing teams are well intended, they usually did not go to school for safety. Their expertise is in other matters. Therefore, the revised process becomes one of helping others to understand what is required or requested, why, and to address value and benefits to the organization. It is not usually as easy as just saying, "we need it because it's a safety requirement."
2. Organizations should not expect or require that every safety improvement project have a positive financial return on investment (ROI) or meet a certain payback period or internal rate of return (IRR) in order to be approved. There can be cases where improvements would have a negative (especially short-term) ROI yet must still be implemented. This could be the case for regulatory compliance issues or cases where there is a high-risk profile.

   Beyond that, there can be problems associated with estimating safety-related *value* reliably. While methods can be and are used to estimate injury or citation cost avoidance, the quality of estimates can vary depending upon project scope. In some cases, it can be quite easy to demonstrate a positive return and even show process improvement and cycle time reduction. In other cases, it can be more difficult. Not because there would be no value in making a proposed change, but because the variables are difficult or impossible to estimate reliably. Will the change mean your organization is not inspected, cited and fined for noncompliance? Will it avoid injuries, and how many, and of what severity, and had injuries occurred, they would have cost … how much? How might you place a value on increased goodwill from employees or on increased employee retention?

It's just not the same as for developing the business case to (figuratively) purchase and install a new machine to make widgets for customer "X," with input costs known, and with a limited range of possible outcomes and profit margins. Organizational leadership should not always expect to analyze project value in the same way, either.

3. It can help OSH staff to communicate about proposed safety improvements in ways most familiar to the project decision-makers and senior executives. Putting the request (business case) in a framework most familiar and relevant to reviewers can go a long way, if the organization has more formal processes.

   Not all organizations "do" have a formal finance function or necessarily have standard processes for project approvals. In smaller organizations, things can sometimes be decided quite easily and relatively informally. The dollar amount needed can impact how formal the process has to be.

Then again, some (many) organizations do have formal processes to approve what they refer to as capital budget requests. In those cases, particular request formats or software might be needed.

Even if a project would not need to be considered a capital project, it could (resources and time permitting) still be good to gather and share certain categories of information as part of a "business case" to make an improvement.

Without spending *too* much time reading your emails or reviewing information through associations or safety software and services providers, you will probably encounter several articles about the *ROI of safety*. The thing is, ROI with respect to safety can mean many different things, and the value that safety programs provide can be sliced and diced many different ways. ROI is just one type of financial metric, and there are several other types of value that can be provided by a safety initiative that are very important but yet not directly financial.

The more broadly one thinks of safety and possible safety benefits, the more difficult it seems to get to describe it. There can be many benefits, financial and intangible. To my knowledge, there is no "universal theorem" approach to describing safety value or safety ROI.

The most solid business cases and calculations of costs and benefits that I have seen have centered around specific projects and specific improvements, e.g., a new material handling system or improved guarding and engineering controls.

You may or may not need to do this, but there are several good ideas you can find online for how to build a good business case. Business case elements that could be provided from the perspective of someone managing safety include things like:

- Executive summary
- Statement of need/high level summary of OSH risk and risk to the organization
- Proposed solution, including time sensitivities and level of urgency
- Summary of the benefits of making recommended changes
- Description of alignment of company goals with the goals of this project
- Risk evaluation data – risk details stemming from current state
- Cost-benefit analysis, including common financial metrics like ROI and payback period, and intangible benefits
- Summary – recommendation of preferred option(s)

There are, no doubt, vendors that are willing to help potential customers estimate the value returned by implementation of solutions they provide. For example, I've recently seen ROI-based marketing literature related to personal protective equipment changes, and related to the implementation of artificial intelligence solutions.

Again, how you approach cost-benefit analysis (sometimes also referred to as benefit-cost analysis) calculations can depend upon many factors. It's not necessarily difficult but more that we can cover here. One metric that everyone talks about is ROI. The basic formula for ROI, in its simplest form, is not difficult:

$$\text{ROI} = (\text{Gains/benefits from investment} - \text{Cost of investment}) \div (\text{Cost of investment})$$

What can take more effort on your part, and which will always need to be case-specific, is categorizing what you need to spend and when, what the benefits will be and when (either positive financial returns or cost avoidance), the time frame that your organization would like considered in the analysis (one year, five years, etc.), and the time value of money if it makes sense to consider it for your case (accounting for interest rates).

There is a template available for free that you might consider for use in the development of a business case for an OSH project, depending on company needs. It can help you with several common business case elements or give you things to think about. It is located on the website of the American Industrial Hygiene Association (AIHA)[6]. You do not need to be a member to access it. It is described as the "AIHA Business Case in EHS Tool."

There are also tools available through the National Safety Council (NSC). Additionally, courses about the "business case" are available online such as through the American Society of Safety Professionals (ASSP).

The bottom line is that there are tools that you can find to give you some specific ideas about how to approach *the value of safety*. I would encourage you to think about overall value, as opposed to purely financial approaches. You can find helpful tools that provide framework for financial calculations as well as describing intangible value. However, the first priority should be to avoid unacceptable OSH risks that need correction to begin with. Building some PtD concepts in your safety program or system can help with that.

Notes

1 Franklin, B. (1776). *Poor Richard's Almanack.*
2 American National Standards Institute. (2023). *Safety of Machinery* (ANSI B11.0–2023). American National Standards Institute.
3 American Society of Safety Professionals. (2021). *Prevention through Design Guidelines for Addressing Occupational Hazards and Risks in Design and Redesign Processes* (ANSI/ASSP Z590.3–2021). American National Standards Institute.
4 A list of approved NRTLs can be found at https://www.osha.gov/nationally-recognized-testing-laboratory-program/current-list-of-nrtls.
5 Note that OSHA had adopted the ANSI B11.1 standard, 1971 version, to create OSHA standard 29CFR1910.217, Mechanical Power Presses. This is different than incorporating specific sections of it by reference. The ANSI B11 standards series has expanded since then in number and scope and are often used in industry, but are not incorporated by reference.
6 https://www.aiha.org/public-resources/healthierworkplaces/healthier-community-resources/apps-and-tools-resource-center.

12 Where to Get OSH Help

Next Steps

In prior chapters, we reviewed where requirements for written occupational safety and health (OSH) programs come from, the value they can add, things to consider during development, a glimpse into the safety market, and some direct tools for program development. Beyond that, I've shared information about other things that, even though not mandated by the Occupational Safety and Health Administration (OSHA) or State Plans, could be important for you to consider. That includes ideas related to program or system implementation and maximizing what you obtain from your efforts. I've also shared a couple of practical implementation examples, and some links to resources for further information.

For many situations, that can be enough to help an OSH program developer put together an action plan and move forward. However, in other cases, staffing, circumstances, and complexity may suggest or require that the organization obtain additional assistance. That assistance could be strategic and broad, or very focused on specific issues.

Context for Need

In Chapter 2, we reviewed sources of requirements for OSH programs, and organizational context. A variety of internal and external factors may drive your organization's OSH program needs, which could stem from things like:

- Regulatory mandates
- Regulatory inspections and citations for violations
- Legal requirements for abatement of issues and/or settlements
- Fatal and nonfatal injury and illness cases
- Significant "near misses"
- Organizational values and wanting to "do the right thing" for employees
- Other business drivers

Those "other" business drivers can include desires to be known for having an OSH program of high quality that offers your clients expectations of high

DOI: 10.4324/9781032704586-12

performance, as well as reliability as a supplier of products or services. It can suggest a level of organizational sophistication and increase trust in your organization. It is also an indication of an organization's values, which can be a positive factor in the development of new opportunities and partnering arrangements.

It can be easy for some third parties to say that everyone developing a safety program should be a degreed safety professional and have a certain number of years of experience. It is true that some organizations have a high-enough risk profile for that to be essential. Sometimes, specific types of OSH issues can be important enough or intense enough to drive that. (See also Chapter 7.)

However, not all organizations are of a size or have the inherent risk or process complexity to feel that is warranted.

The bottom line is that while many organizations can provide internal resources to develop their own OSHA-mandated programs, others may need assistance to build a "spot-on" OSH program or system, or to have some level of comfort that critical issues have been or will be addressed.

Presumed Staff Experience

While not every organization will have a full-time, dedicated safety professional as an internal resource, it is still the case that a certain skill set is needed to set up and run an effective OSH program. it is important to know that "not just anyone" can be assigned the task of developing an OSH program and be expected to succeed at it.

Whoever is in that role should optimally have some project management experience and ability to "get things done" in the organization, have an understanding of the organization, and some understanding of OSH frameworks and potential issues. Creating written programs does not require that the leader of that effort be a technical expert in all matters, but resources do need to be available.

The level and type of resources needed, either for OSH program development or subsequent OSH management, can depend on several things.

Levels of Need

When creating a new OSH program, keep in mind that the process can entail not only setting up the framework for the program but also creating and facilitating the processes for "how" things will work. That can include the underlying forms or checklists and similar items that might be needed to engineer desired processes. Therefore, some need for assistance may stem from "writing" things, but other types of support could be needed, too.

With that being the case, the level of burden in creating a new program is driven by numerous factors. Examples of things that can impact OSH burden are listed below. (Some of these pertain to organizations that have not yet started operations, and others pertain to those that are already underway.)

- The number of processes regulated under OSHA/State Plans and other agencies (e.g., environmental regulations with OSH components)
- The complexity of processes and things for which OSHA-specified actions and documentation will be required
- The number of employees who will be exposed to particular hazards
- The number and inherent level of risks to the organization
- The extent to which an occupational health and safety management system (OHSMS) approach is desired
- The processes that must be created to conform to OHSMS requirements (as desired)
- The need to conduct evaluations of possible exposures to hazardous materials or ergonomic risks
- The amount of coordination required in order to make decisions (e.g., with maintenance, operations, controls staff, human resources)
- The number of technical details that might need to be worked through before being able to write detailed procedures (such as for equipment-specific energy control procedures)
- The extent and complexity of training requirements, including determination of training needs, development, methods of provision, tracking of training
- Other underlying processes and forms providing details and methods of documentation
- Organizational culture and level of buy-in to safety concepts and requirements
- The anticipated presence of high- or very high-risk tasks with serious injury and fatality potential
- The frequency and severity of injuries
- The level of other (non-safety) assigned duties, and the predictability of competing time demands
- The OSHA-related education and aptitudes of the OSH program developer

As a general thought, it is likely that …

- the higher an organization's need for and reliance upon safety controls,
- the higher the complexity of inherent operations and hazards,
- the poorer the safety culture,
- the higher the possible public and regulatory interface, and
- the greater the possible impact of something going wrong,

… the higher the need for assistance.

Support Options

There are several ways to obtain different forms of support and assistance. Common ways include receiving free assistance through OSHA consultation programs, working with state business development related programs, industry and trade associations, and other types of consultants. Here are some examples.

OSHA Consultation Programs and Economic Development Grants

Chapter 3 discussed on-site consultation programs available in each state, focusing on small and medium-sized employers. That chapter provides a URL to onsite consultation programs by state. There can be other forms of consultation programs within a state as well. Formats and methods of working with employers can vary. Some consultation programs cover broader segments of employers than others. Your organization may or may not be eligible for help, depending upon the type of work your organization does or will perform, and because prioritization may be given to employers where exposures to hazards are already occurring. However, even if in a lower-hazard industry, you still might be able to obtain some help – but "it depends." Where help through those programs may be obtained, it can be excellent. It's worth looking into.

Some states offer training programs intended to help organizations and workers to understand OSH obligations and issues, and otherwise offer training opportunities that can be helpful to (especially) small and medium enterprises (SMEs).

Some states also have various forms of economic development programs that provide training, which can sometimes include OSH topics. If your organization is eligible, that training assistance could be quite useful.

Industry and Association Support

Some industry groups and associations have strong health and safety committees and services for their members. That can sometimes include networking, training, provision of guidance documents, safety mentoring programs, and auditing. They can be very good, since they've often worked through significant OSH issues that impact members of their communities, sometimes including lobbying or advocating for their members regarding critical issues, in Washington, DC. Consider investigating association resources that might be available to you.

Consultants

There are many OSH consulting firms and independent OSH consultants that may be of assistance. There are many different areas of focus and specialty.

Consultants may be able to advise you strategically and confidentially in ways that other organizations cannot. The next few paragraphs share some perspectives on potential value they can provide.

Value that Can Be Added

One critical thing that consultants can do is give early input to you (before too much time or "other" activity passes by) that can help you ensure and/or feel comfortable that your organization has its bases covered and is moving in the right direction.

They can also work with you as a partner to discuss your plans in both broad and narrow ways. For example, they can help suggest staff credentials, job descriptions and interview processes, reporting structures, management strategy, and what they've seen work or not work elsewhere. Importantly, they can help your organization to make sure that critical risks and needs are identified, provide opinions on OSHA requirements, and suggest or assist with best practices for OSH management.

Beyond that, you could ask consultants to, at the right time, assist with program or system development, and provide in-depth technical or other specialized evaluations.

Different Types and Areas of Focus

Consulting firms and independent OSH consultants can provide a mix of services. However, for the sake of this discussion, some types of consulting services and areas of focus can be compartmentalized into the following categories.

Generalists: They address a wide range of OSH issues, often with certain areas of focus, based on interests and experience. They often have extensive resources available and may work with a network of specialists. They are good at many things, but may need to refer highly specialized issues to other technical experts.

OHSMS specialists: These consultants focus on management system consensus standards that pertain to OSH, like the ANSI/ASSP Z10.0 standard and the ISO 45001 standard. The focus of OHSMSs is more on OSH processes and making sure that the right set of interrelated processes are in place than the specific requirements of OSHA standards. While OHSMS standards expect compliance with OSHA requirements, OHSMS implementation is intended to help organizations move beyond *just* compliance. OHSMS specialists can help organization to develop a system comprised of policy, programs, and procedures, and related processes, to manage OSH with systems and risk-based thinking.

Combined/integrated environment, health, and safety (EHS) consultants: These are firms that provide services for both the OSH and environmental subject areas in an integrated way. However, firms may still divide their staff into primary areas of specialty. Like the OSH field, the environmental management field is very broad. As a general note, if you evaluate consulting firms to provide services to your organization, you may notice that some are more focused either on providing environmental-related services, or on providing OSH services. Those with a strong environmental focus may not be the strongest in OSH concepts and methods, and vice versa. It's natural that focus may be stronger in one technical area than another; pick appropriately based on primary issues likely to face your organization.

Industry-focused: Many consulting firms focus on, or only work within, particular industries. They may also want to work only with certain types of clients. For example, consultants may focus on automotive, construction, healthcare, metal casting, oil and gas, pharmaceutical, retail and warehousing, or transportation.

There are also consultants that focus on particular "types" of hazards.

Machine safety: These firms/consultants specialize in machinery and equipment–related assessments, regulations and consensus standards, and related issues. Their focus is often on the highly technical aspects of machine technology, safety-related software and components, machine guarding, safety controls, control reliability, and other matters. As such, their work can be closely related to OSHA's machine guarding and energy control standards.

Occupational hygiene: Organizations that provide occupational hygiene (often also referred to as industrial hygiene) services often do so across multiple industries, although there are some that focus on a limited number. They can often perform many different types of assessments intended to help organizations comply with occupational exposure-based regulations and to meet guidelines that organizations may choose to follow. They may often be the go-to consultants for programs related to chemical safety management, exposure assessments, respiratory protection programs, confined space programs, biological monitoring, and others. Many occupational hygiene consultants also provide environmental assessment and management-related services.

Process safety management: This subject area focuses on OSHA's process safety management (PSM) standard located at 29 CFR 1910.119. Such consultants can, as qualified, assist with highly technical program development and application of engineering knowledge, such as for determination of whether or not processes meet RAGAGEP (recognized and generally accepted good engineering practices) criteria. These firms are often very strong in engineering and with a chemical industry focus.

Selection Considerations

No doubt you can see that there are many types of OSH consultants one could hire. Some might be better choices for you than others, depending on factors like your level of OSH background, what your organization hopes to achieve, the industry you work in, and the firm's or consultant's expertise.

Questions to ask before your organization makes arrangements for consulting services can include:

- How would you like the firm/consultant to engage with you? As a partner or on a more limited basis? Strategic or hands-on?
- Does the firm/consultant have relevant experience with your organization's nature of services or operations, and critical issues that you may anticipate?

- Who are the critical players, and how did they obtain their relevant expertise?
- Does the firm/consultant have relevant knowledge of applicable OSH regulations in your industry, and have they satisfied the regulators that you may interact with?
- Are the firm's/consultant's typical projects aligned with the type and level of detail that you are likely to need?
- Are the firm's/consultant's typical projects either much larger in scope, or much smaller in scope, than your project?
- If needed, can the firm/consultant shift to other types of service and assessments as the project evolves?
- Who will perform the work?
- Would the firm/consultant be a good cultural fit?
- Would your firm be likely to receive the level of attention it needs, in the time frame that it needs? How do they describe their response time?
- Is the firm/consultant likely to communicate with you in a manner that matches your background and needs?

Putting It All Together

This book has described requirements, facts, and background information that can be of use to you as you develop and implement OSHA-mandated written programs. It has also shared other relevant information and tools that you may want to know about and consider as you work to optimize the return you obtain for your efforts.

It may be appropriate to receive some external assistance. Several sources of help are likely to be available to you – plan carefully.

Finally, it was my intent to provide practical and helpful information to you, from an insider's perspective. I hope that it will be of value to you now and as a reference in the future.

I would be happy to receive notes from readers describing success stories and experiences as you move forward, through Sixth Sense Safety Solutions (https://www.sixthsensesafetysolutions.com).

Index

Note: **Bold** page numbers refer to tables, *italic* page numbers refer to figures and page numbers followed by 'n' refer to notes.

For Product Safety Concerns and Information please contact our EU
representative GPSR@taylorandfrancis.com
Taylor & Francis Verlag GmbH, Kaufingerstraße 24, 80331 München, Germany

www.ingramcontent.com/pod-product-compliance
Ingram Content Group UK Ltd.
Pitfield, Milton Keynes, MK11 3LW, UK
UKHW020005200926
475702UK00015B/2479

* 9 7 8 1 0 3 2 7 0 4 4 8 7 *